The Earth: A Very Short Introduction

'This is a splendid introduction to the Earth. There is enough detail here to get you really into an understanding of how our dynamic planet "works", based in its history and structure. . . . Writing with great clarity and obvious enthusiasm Martin Redfern has managed to capture the excitement of the new discoveries in the Earth Sciences. I found it hard to put down!'
Aubrey Manning, University of Edinburgh

'An excellent introduction to modern Earth Sciences Martin Redfern's authoritative and timely account will be invaluable to secondary school students, entry level undergraduates, as well as all those interested in modern geoscience.'
Richard Corfield, author of *Architects of Eternity* and *The Silent Landscape*

VERY SHORT INTRODUCTIONS are for anyone wanting a stimulating and accessible way into a new subject. They are written by experts, and have been translated into more than 45 different languages.

The series began in 1995, and now covers a wide variety of topics in every discipline. The VSI library now contains over 500 volumes—a Very Short Introduction to everything from Psychology and Philosophy of Science to American History and Relativity—and continues to grow in every subject area.

Titles in the series include the following:

Martin Redfern

THE EARTH

A Very Short Introduction

OXFORD
UNIVERSITY PRESS

Great Clarendon Street, Oxford OX2 6DP

Oxford University Press is a department of the University of Oxford.
It furthers the University's objective of excellence in research, scholarship,
and education by publishing worldwide in

Oxford New York

Auckland Bangkok Buenos Aires Cape Town Chennai
Dar es Salaam Delhi Hong Kong Istanbul Karachi Kolkata
Kuala Lumpur Madrid Melbourne Mexico City Mumbai Nairobi
São Paulo Shanghai Taipei Tokyo Toronto

Oxford is a registered trade mark of Oxford University Press
in the UK and in certain other countries

Published in the United States
by Oxford University Press Inc., New York

British Library Cataloguing in Publication Data

Data available

Library of Congress Cataloging in Publication Data

Data available

ISBN 978-0-19-280307-8

21

Typeset by RefineCatch Ltd, Bungay, Suffolk

Printed and bound by
CPI Group (UK) Ltd, Croydon, CR0 4YY

Acknowledgements

The author would like to thank: Arlene Judith Klotzko for introductions without which this book would never have been written; Shelley Cox for initial enthusiasm; Emma Simmons for continued patience; David Mann for instant cartoons; Pauline Newman and Paul Davies for helpful comments; Marian and Edmund Redfern for nurturing my enthusiasm and reading the results; Robin Redfern for beavering away; the un-named readers who have kept me precise; and the countless geologists who have shared with me their time and enthusiasm.

Contents

List of illustrations

The publisher and the author apologize for any errors or omissions in the above list. If contacted they will be pleased to rectify these at the earliest opportunity.

1. Planet Earth, as seen from *Apollo 17*, December 1972.

Chapter 1
Dynamic planet

Once a photograph of the Earth, taken from the outside, is available,
a new idea as powerful as any in history will be let loose.

Sir Fred Hoyle, 1948

How can you put a big round planet in a small flat book? It is not an
easy fit, but there could be two broadly different ways of attempting
it. One is the bottom-up approach of geology: essentially, looking at
the rocks. For centuries, geologists have scurried around on the
surface of our planet with their little hammers examining the
different rock types and the mineral grains which make them up.
With eye and microscope, electron probe and mass spectrometer,
they have reduced the planet's crust to its component parts. Then
they have mapped out how the different rock types relate to one
another and, through theory, observation, and experiment they
have worked out how they might have got there. It has been a huge
undertaking and one that has brought deep insights. Collectively,
the efforts of all those geologists have built a giant edifice on which
future earth scientists can stand. It's as a result of this bottom-up
approach that I can write this book. But it is not the approach I will
use. This is not a guide to rocks and minerals and geological map-
making. It is a portrait of a planet.

The new view on our old planet is the top-down approach of what
has come to be known as Earth systems science. It looks at the

Earth as a whole and not just frozen in time in the moment we call now. Taken over the deep time of geology we begin to see our planet as a dynamic system, a series of processes and cycles. We can begin to understand what makes it tick.

The view from above

The prediction above was made by astronomer Sir Fred Hoyle in 1948, a decade before the dawn of space flight. When unmanned rockets took the first pictures of the Earth from outside, and when the first generation of astronauts saw for themselves our world in its entirety, the prediction came true. It's not that those first views told us much we didn't already know about the Earth, but they gave us an icon. And to many of the astronauts who witnessed the view first-hand, it gave an emotional experience of the beauty and seeming fragility of our world that has lived with them ever since. It is perhaps no coincidence that Earth sciences were undergoing their own revolution at the same time. The concept of plate tectonics was at last gaining acceptance, 50 years after Alfred Wegener originally suggested it. Exploration of the ocean floor revealed that it was spreading out from a system of mid-ocean ridges. It had to be going somewhere, forcing continents apart or into one another. The unimaginable masses of continent-sized plates of rock were on the move in an elaborate and ancient waltz.

It was around the same time, and with the same icon of that small blue jewel we call Earth floating in the blackness of space, that a global environmental movement began to form, a mixture of those with a sentimental attachment to endangered species and rainforests and scientists taking on board a new view of complex, interacting ecological systems. Today, most university departments and research groups use the term 'Earth sciences' rather than geology, recognizing a broadening of the discipline beyond the study of rocks. The term 'Earth systems' is becoming widespread, recognizing the inter-related, dynamic nature of processes that include not only the solid, rocky Earth but its oceans, the fragile veil of its atmosphere, and the thin film of life on its surface as well. It is as if our world were an onion; a series of concentric spheres, from magnetosphere and atmosphere, through biosphere and hydrosphere, to the layers of the solid earth. Not all are spherical and some are much less substantial than others, but each manages to persist in a delicate equilibrium. Each component of such a system is seen not as something fixed and unchanging but more like a fountain; maintaining its overall structure perhaps, but constantly changing as material and energy pass through it.

If rocks could talk

Rocks and stones are not the most forthcoming of storytellers. They have a tendency to sit there gathering moss, only rolling when pushed. But geologists have ways of making them talk. They can hit them and slice them; squeeze them, squash them, strain and stress them until they crack – sometimes quite literally. If you know how to look at them, rocks can tell you their history. There is the recent history of the rock on the surface: how it has been weathered and eroded; the tell-tale scars of wind, water, and ice. There can be deeper scars that record periods of heat and pressure and deformation when the rock was buried. Where these changes are extreme the rocks are known as metamorphic. Then there are clues to the origin of the rocks. Some show signs of having once been molten and pushed up from deep within the Earth to erupt out of volcanoes or to intrude into pre-existing rocks. These are the igneous rocks. The size of mineral grains within them can reveal how quickly they were cooled. A large mass of granite cools slowly so that crystals in it are large. Volcanic basalt solidifies rapidly and so is fine-grained. Rocks can be made of the ground-down remains of previous rocks. Here, the size of the fragments tends to reflect the energy of the environment that laid them down: from fine shale and mudstone deposited in still water, through sandstones to coarse conglomerates washed down by raging torrents. Others, such as chalk and limestone, are chemical deposits accumulated as living systems took carbon dioxide from the atmosphere and precipitated it in sea water, turning, as it were, the sky into stone.

Even individual mineral grains have their story to tell. Mineralogists can strip them apart atom by atom in mass spectrometers so sensitive that they can reveal different ratios of isotopes (different atomic forms of the same elements) even among trace constituents. Sometimes these can help date the grains so that we know if they came out of still more ancient rock. They can also reveal the stages of growth of a crystal, for example of diamond, as it passes through the Earth's mantle. In the case of isotopes of carbon and oxygen in minerals derived from marine organisms, it's even possible to estimate the temperature of the sea and the global climate when they formed.

Other worlds

The trouble with the Earth is that it is the only one we've got. We can only see it as it is today, and we can't tell if it is here simply due to some happy accident. That's why Earth scientists are taking a renewed interest in astronomy. Powerful new telescopes sensitive to infrared and sub-millimetre wavelength radiation can stare deep into star-forming regions to see what may have happened when our own solar system was born. Around some of the young stars they have revealed dusty haloes known as proto-planetary discs, perhaps new solar systems in formation. But the search for fully formed Earth-like planets is more difficult. To see directly such a planet in orbit around a distant star would be like trying to spot a small moth close to a powerful searchlight. But indirect methods have led to the discovery of planets in recent years, mainly by detecting tiny wobbles in the motions of the parent stars due to gravitational effects. The clearest effects and therefore the first discovered seem to be due to planets far bigger than Jupiter orbiting far closer to their stars than the Earth is to the Sun. So they could hardly be termed Earth-like. But evidence is beginning to accumulate for solar systems more like our own, with multiple planets. Though small and hospitable planets like Earth will be hard to detect.

To see such planets directly would take telescopes in space that we can scarcely dream of. There are ambitious plans underway in both the USA and Europe for a network of linked infrared telescopes. Each would have to be far bigger than the Hubble Space Telescope, and four or five of them would have to fly in close formation to combine their signals to resolve the planet. They would have to be as far out as Jupiter to get beyond the dusty infrared glow of our own planetary system. But then, they might be able to detect vital signs in distant planetary atmospheres and, in particular, they might detect ozone. That would imply Earth-like conditions of climate and chemistry plus the existence of free oxygen, something which, as far as we know, can only be maintained by life.

Signs of life

In February 1990, on its way out of the solar system after encounters with Jupiter and Saturn, the Voyager I probe beamed back the first image of our entire solar system as it might appear to visitors from another star. The picture is dominated by a single bright star, our Sun, seen from 6 billion kilometres away, 40 times the distance from which we are used to seeing it. The planets are scarcely visible. The Earth itself is smaller than one picture element in Voyager's camera, its faint light caught in what looks like a sunbeam. This is our whole world, seemingly just a speck of dust. But to any alien visitor with the right instruments, that tiny blue world would immediately attract attention. Unlike the giant stormy gas bags of the outer planets, cold, dry Mars, or the acid steam-bath of Venus, the Earth has everything just right. Water exists in all three phases – liquid, ice, and steam. The atmospheric composition is not that of a dead world that has reached equilibrium but one that is active and must be constantly renewed. There is oxygen, ozone, and traces of hydrocarbons; things that would not exist together for long if they were not constantly renewed by living processes. This alone would attract the attention of our alien visitors, even if they could not detect the constant babble of our communications, radio and television.

Magnetic bubble

Geophysics goes way above our heads. I don't mean by that that it is incomprehensible but that the physical influence of our planet extends far above its solid surface, way out into what we regard as empty space. But it is not empty. We live in a series of bubbles nested like Russian dolls one within another. The Earth's sphere of influence lies within the greater bubble dominated by our Sun. That in turn lies within overlapping bubbles blown by the expanding debris of exploding stars or supernovae, long, long ago. They are all within our Milky Way galaxy, which is in turn a member of a super-cluster of galaxies within the known universe, which itself may be a bubble in a quantum foam of worlds.

The Earth's atmosphere and magnetic field shield us, for the most part, from the radiation hazards from space. Without this protection, life on the Earth's surface would be threatened by solar ultraviolet and X-rays as well as cosmic rays, high-energy particles from violent events throughout the galaxy. There is also a permanent gale of particles, mostly hydrogen nuclei or protons, blowing outwards from the Sun. This solar wind speeds past the Earth at typical velocities of around 400 kilometres per second, and goes three times faster during a solar storm. It extends for billions of kilometres out into space, beyond all the planets and maybe beyond the orbits of comets, which reach out many thousands of times further from the Sun than does the Earth. The solar wind is very tenuous but it is sufficient to blow out the tails of comets as they come closer in to the heart of the solar system, so the tails always point away from the Sun. It also features in imaginative proposals for propelling spacecraft with vast gossamer-thin solar sails.

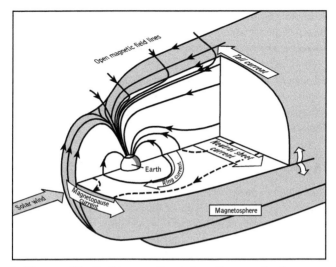

2. Diagram of the Earth's magnetic envelope, the magnetosphere, swept back into a comet-like structure by the solar wind. Arrows show the directions of electrical currents.

The Earth is sheltered from the solar wind by its magnetic field, the magnetosphere. Because the solar wind is electrically charged it represents an electrical current, which cannot cross magnetic field lines. Instead, it compresses the Earth's magnetosphere on the sunward side, like the bow wave of a ship at sea, and stretches it out into a long tail down-wind which reaches almost as far as the orbit of the Moon. Charged particles caught within the magnetosphere build up in belts between the field lines where they are forced to spiral, generating radiation. These radiation belts were first spotted in 1958 when James van Allen flew the first Geiger counter in space on board the American Explorer 1 satellite. They are areas to be avoided by spacecraft hoping for a long life and would be lethal to unprotected astronauts.

Where the Earth's magnetic field lines dive down towards the poles, solar wind particles can enter the atmosphere, sending atoms ricocheting downwards to produce spectacular auroral displays. At the top of the atmosphere, the hydrogen ions of the solar wind itself produce a pink haze. Lower down, oxygen ions produce a ruby-red glow, while nitrogen ions in the stratosphere cause violet blue and red auroras. Occasionally, magnetic field lines in the solar wind are forced close to those of the Earth, causing them to reconnect, often with spectacular releases of energy which extend the auroral displays.

The fragile veil

There's no clearly defined height that marks the top of the atmosphere; 260 kilometres above the ground, in low Earth orbit where the space shuttle flies, you're above almost all the air and the pressure is a billion times less than it is on the ground. But there are still about a billion atoms in a cubic centimetre up there, and they are hot and electrically charged and hence can have a corrosive effect on space vehicles. At times of maximum solar activity, the atmosphere expands slightly, exerting more frictional drag on low spacecraft, which have to be boosted up to stay in orbit. The upper atmosphere, above 80 kilometres, is sometimes known as the thermosphere because it is so hot, even though it is so rarefied that you would not burn your skin on it.

This region of the atmosphere also absorbs dangerous X-rays and some of the ultraviolet radiation from the Sun. As a result, many atoms become 'ionized', that is they lose an electron. For this reason, the thermosphere is also called the ionosphere. Because the ionosphere is electrically conducting, it will reflect certain frequencies of radio waves, making it possible for short-wave radio transmissions to be heard around the world, well over the horizon from the transmitter.

Even a mere 20 kilometres up, below the thermosphere, the mesosphere, and most of the stratosphere, we are still above 90% of the air in the atmosphere. It is at around this height that we encounter the tenuous ozone layer, molecules containing three oxygen atoms. Ozone forms when ordinary oxygen molecules of two atoms are split by solar radiation and some recombine in threes. Ozone is a highly effective sunscreen for the planet. If all the ozone in the Earth's atmosphere were concentrated at ground level, it would form a layer only about three millimetres thick. But it still filters out virtually all of the most dangerous short-wave UV C radiation from the Sun and most of the medium-wavelength UV B rays as well. Thus it protects life from sunburn and skin cancer. The ozone layer has been severely depleted by chemicals such as CFCs (chlorofluorocarbons) released by human activity, leading to a generalized thinning of the layer and more specific holes over polar regions in the still, cold air of spring. International agreements have slowed the release of CFCs and the ozone layer should recover, but the chemicals are long-lived and it will be some time yet before it does.

Circles and cycles

It is in the lowest 15 kilometres of the atmosphere, the troposphere, that most of the action takes place. This is where weather happens. It's where clouds form and disperse and where winds blow, transferring heat and moisture around the planet. In a dynamic planet, everything seems to go round in circles, flows of energy. And here, close to the surface, these cycles are driven by solar power. There are the obvious cycles of day and night as the Earth spins on its axis and the ground alternately heats and cools, and the annual cycle of the seasons as the Earth orbits the Sun, presenting first more of one hemisphere then more of the other to the sunshine. But there are longer cycles too, such as the wobble of the Earth's axis over tens of thousands of years.

Just as the Earth orbits the Sun, so the Moon orbits the Earth. It takes about 28 days to complete an orbit, giving us our months. As the Earth spins on its axis, the Moon's gravity pulls a bulge in the oceans around the planet, creating tides. This also acts as a brake on the rotation of the Earth, slowing down the days. Daily growth bands in fossil corals 400 million years old suggest that their days were several hours shorter than our own.

The Moon helps to stabilize the orbit of the Earth and hence the climate. But there are far longer cycles at work too. The Earth's orbit around the Sun is not a perfect circle, but an ellipse, with the Sun at one focus. Hence the distance of the Earth from the Sun varies during its orbit. In addition, the degree of variation itself changes over a 95,800-year period. Also the Earth's rotation axis slowly wobbles or precesses like a spinning top off balance. Over a period of 21,700 years the planet's axis traces out a complete cone. At present, the Earth is nearest to the Sun during the northern hemisphere winter. The inclination of the Earth's spin axis with that of its orbit around the Sun (the obliquity) also changes on a 41,000-year period. These so-called Milankovich cycles add up over tens or hundreds of thousands of years to affect climate. They have been blamed for such phenomena as the ice ages that have affected the Earth over the last three and a quarter million years. But the reality is probably even more complex, with their effects amplified or reduced by factors such as ocean circulation, cloud cover, atmospheric composition, volcanic aerosols, the weathering of rocks, biological productivity, and so on.

Solar cycles

Cycles of change are not restricted to the Earth. The Sun can change too. Over its 5-billion-year history, the Sun has been getting progressively warmer. Surface temperatures on Earth have remained more constant, however, as levels of greenhouse gases have fallen over the same time. This was largely due to the effects of life, as plants and algae consumed carbon dioxide that acted like a blanket to keep the young Earth warm. There have been other solar variations as well. There is a regular solar cycle of 11 years which sees a rise and fall in sunspot activity, in turn reflecting the cycle of solar magnetic activity which produces storms and the solar wind. Other Sun-like stars seem to spend about a third of their time free of sunspots, a state called a Maunder minimum. That happened to our Sun between 1645 and 1715 AD. Solar power only dropped by about 0.5%, but this was enough to plunge northern Europe into what has become known as the Little Ice Age, with a series of particularly severe winters. The River Thames in London froze over, and markets and frost fairs were held on it.

Hot air

The Sun distributes its warmth unevenly, warming up equatorial regions the most. As the air warms it tries to expand, increasing atmospheric pressure. To try to restore equilibrium, winds begin to blow and the air circulates. Whilst all this goes on, the Earth continues to rotate, giving the air angular momentum. That is greatest at the equator and results in the so-called Coriolis effect. The atmosphere is not firmly coupled to the solid planet, so, as winds blow away from the equator, they have a momentum that is independent of the rotating surface beneath. This means that, relative to the surface, the winds curve to the right in the northern hemisphere and left in the south. This leads to rotating systems of high and low air pressure, the weather systems that bring us rain or sunshine.

Land masses and mountain ranges influence the circulation of heat and moisture too. Until the Himalayan mountain range began to rise, there was no Indian monsoon, for example. And most importantly, the oceans play a huge part in storing heat and transporting it around the globe. The top 2 metres of the ocean have the same heat capacity as the entire atmosphere. At the same time, heat circulates in ocean currents. But currents on the surface are only half the picture. A good example is the Gulf Stream in the North Atlantic. That carries warm water north and east from the Gulf of Mexico and is one of the reasons why the climate of northwest Europe is much milder in winter than that of northeast America. As the warm water heads north, some evaporates into the clouds, which always seem to fall on British holidaymakers. The remaining surface waters in the ocean cool and become progressively more salty. As a result, they also become denser and eventually sink down to flow back south in the deep Atlantic, completing the conveyor belt of the ocean circulation.

Sudden freeze

About 11,000 years ago, the Earth was emerging from the last Ice Age. Ice was melting, sea levels were rising, and the climate was getting generally warmer. Then, suddenly, in the space of a few years, it turned cold again. The change was particularly marked in Ireland, where pollen in sediment cores shows that the vegetation suddenly reverted from temperate woodland back to tundra dominated by a little plant called the Dryas. Wally Broecker of the Lamont Doherty Geological Observatory has worked out what may have happened. As the ice sheet over North America receded, a vast lake of fresh meltwater, far bigger than the present Great Lakes, built up in central Canada. At first, it drained over a great ridge of rock into the Mississippi. As the ice receded, it suddenly opened a far lower passage down the St Lawrence River to the east. The vast lake of cold fresh water drained into the North Atlantic almost instantaneously. So much water was involved that it caused an immediate sea level rise of 30 metres. It diluted the salty surface water of the North Atlantic and put a virtual stop to the conveyor belt of ocean circulation. Thus there was no warming current into the North Atlantic and Arctic conditions returned. A thousand years later, the ocean circulation resumed as quickly as it had vanished and a temperate climate returned.

The North Atlantic deep water, together with cold bottom water from the Antarctic, finds its way at depth as far as the Indian and

Pacific oceans. The deep current continues into the North Pacific, slowly accumulating nutrients as it goes, before it rises again to the surface.

Global greenhouse

Some of the gases in the Earth's atmosphere act rather like the glass of a greenhouse, letting sunlight in to warm the ground but then preventing the resulting infrared heat radiation from escaping. Were it not for the greenhouse effect, average global temperatures would be around 15 degrees Celsius lower than they are, making life almost impossible. The principal greenhouse gas is carbon dioxide but others, including methane, play an important role. So does water vapour, an effect that is sometimes forgotten. Over hundreds of millions of years, an approximate balance has been struck with plants removing carbon dioxide from the atmosphere through photosynthesis and animals returning it by respiration. Vast quantities of carbon have been buried in sediments such as limestone, chalk, and coal. Volcanic eruptions have released carbon from inside the Earth.

In recent years, concern has grown over what should be termed the enhanced greenhouse effect, the very significant rise in greenhouse gas levels in the atmosphere resulting from human activity. The burning of fossil fuels such as coal and oil are prime culprits, but so are agricultural practices which produce methane, and deforestation which releases carbon dioxide from timber and soils as well as reducing the plant cover to absorb it again. Climate models suggest that these activities could result in a global temperature rise of several degrees over the next century, accompanied by greater extremes in weather and a possible sea level rise.

Climate change

The steady annual rise in carbon dioxide levels has been carefully recorded from an isolated mountain peak in Hawaii since 1958. More than 130 years of meticulous weather readings around the world confirm an average global warming of about half a degree, with the effect particularly pronounced over the last 30 years. But natural climate records go back much, much further. Tree rings record periods of drought and severe frosts, as well as the frequency of wild fires, over their lifespan. Extrapolating overlapping sequences in preserved timber can reveal climate conditions back to 50,000 years ago. Coral growth rings reveal sea surface temperatures over a similar span. Pollen grains in sediments record shifts in vegetation patterns over 7 million years. Landscapes reveal past glaciation and changes in sea level over billions of years. But some of the best records come from cores drilled from ice and from ocean sediments. Ice cores reveal not only rates of snow accumulation and trapped volcanic dust, but bubbles in the ice represent samples of the ancient atmosphere trapped in the snow. Isotopes of hydrogen, carbon, and oxygen can also indicate global temperature at the time. The ice record from Antarctica and Greenland now goes back over 400,000 years. Marine sediments all around the planet have been sampled by the ocean drilling programme and can carry records up to 180 million years old. Isotope ratios in microfossils trapped in the sediments can reveal temperature, salinity, atmospheric carbon dioxide levels, ocean circulation, and the extent of polar ice caps. All these different records reveal that climate change is a fact of life and that long periods in the past have been considerably warmer than the climate we experience today.

Web of life

The most insubstantial of the Earth's layers has had perhaps the most profound effect on the planet: life. Without life, the Earth might be a runaway greenhouse world like Venus, or possibly a cold desert like Mars. There would certainly not be the temperate climate and oxygen-rich atmosphere in which we flourish. We've already heard how the first algae kept pace with the warming Sun by eating the carbon dioxide blanket that insulated the young Earth. The independent scientist James Lovelock suggests that such feedback mechanisms have managed the terrestrial climate for more than 2 billion years. He uses the term Gaia, after the ancient Greek earth goddess, as a name for this system. He does not pretend that there is anything conscious or deliberate about this control; Gaia does not have divine powers. But life, principally in the form of bacteria and algae, does play a key role in the homeostatic process that keeps the planet habitable. A simple computer model called 'Daisyworld' shows how two or more competing species can set up a negative feedback system that controls the environment within habitable limits. Lovelock suspects that the global system on Earth will adapt as human activity enhances the greenhouse effect, even if the adaptations are not favourable to human life.

The carbon cycle

Carbon is forever moving around. Each year, roughly 128 billion tonnes is released as carbon dioxide into the atmosphere by processes on land, and nearly as much is immediately absorbed again by plants and by the weathering of silicate rocks. At sea, the figures are comparable, though slightly more goes in than comes out. The system would be more or less in balance were it not for volcanic emissions and the 5 billion tonnes released each year by burning fossil fuels. The total amount of carbon held in the atmosphere is quite small – just 740 million tonnes, only slightly more than that held in plants and animals on land and slightly less than that held by living things in the ocean. By comparison, the amount of carbon stored in solution in the oceans is vast at 34 billion tonnes, and the amount stored in sediments is 2,000 times greater still. So the physical processes of solution and precipitation may be even more important in the carbon cycle than the biological ones. But life seems to hold some key cards. Carbon incorporated by phytoplankton would be released back to sea water and hence the atmosphere very quickly were it not for the physical properties of copepod faecal pellets. These tiny planktonic animals excrete their waste in small, dense pellets which can slowly sink into the deep ocean, removing them, at least temporarily, from the cycle.

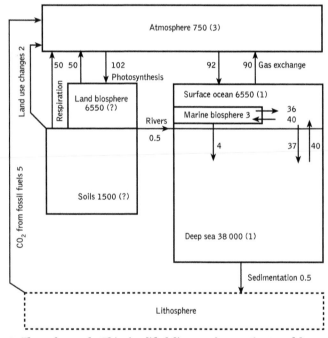

3. The carbon cycle. This simplified diagram shows estimates of the amount of carbon (in billions of tonnes) stored in the atmosphere, oceans, and land. Figures by the arrows show the annual fluxes between stores, figures in brackets the annual net rise. Though small compared with most other fluxes, the input from burning fossil fuels seems enough to upset the balance.

Almost an onion

The interior of the Earth is rather like an onion, made up of a series of concentric shells or layers. On the top is a crust, averaging 7 kilometres thick under the ocean and 35 kilometres thick in continents. That sits on the hard, rocky lithosphere at the top of the mantle, and below it is the softer asthenosphere. The upper mantle extends to a depth of about 670 kilometres, the lower mantle goes down to 2,900 kilometres. Below a thin transition layer comes the liquid outer core of molten iron and a solid iron inner core about the size of the Moon. But it is not a perfect onion. There are horizontal differences within layers, variations in thickness of layers, and, we now know, continuous exchange of material between layers. Where our planet departs from the perfect onion model is where most off the interest and excitement in modern geophysics

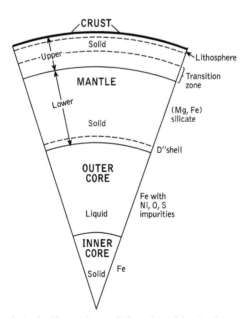

4. The main 'onion' layers in a radial section of the Earth.

lies, and where we can find the clues to the processes that drive the system.

Lava lamps

Do you remember those lava lamps of the 1960s and numerous later revivals? They make a good model for the processes at work within the Earth. Whilst they are switched off, a layer of red gloop sits at the bottom of a layer of transparent oil. But turn on the lamp and the filament in its base warms the red gloop so it expands, becomes less dense, and begins to rise in elongated lumps to the top of the oil. When it has cooled sufficiently, it sinks back down. So it is in the Earth's mantle. Heat from radioactive decay and from the Earth's core drives a sort of heat engine in which the not-quite-solid rock of the mantle slowly circulates over billions of years. It is this circulation that drives plate tectonics, causes the continents to drift, and triggers volcanoes and earthquakes.

The rock cycle

At the surface the results of that heat engine beneath our feet and the solar furnace above our heads meet and drive the rocks full circle. Mountain ranges lifted high by mantle circulation and continental collisions are weathered down by solar-powered wind, rain, and snow. Chemical processes are at work as well. Oxidation by the atmosphere and chemical dissolution by acids from living organisms and dissolved gases help to break down the rocks. Large quantities of carbon dioxide can dissolve in rainwater to make a weak acid that causes chemical weathering, turning silicate minerals into clay. These remains are washed back down to estuaries and oceans where they form new sediments, eventually to be scooped up into new mountain ranges or carried back down into the mantle for deep recycling. The whole process is lubricated by water incorporated into the crystal structure of minerals. This rock cycle was first suggested in the 18th century by James Hutton, but then he had no idea of the depths and the time scales over which it occurs.

So far we have glimpsed just the surface of our amazing planet. Now we will dig deeper into the rocks and into time.

Planetary data

Equatorial diameter	12,756 km
Volume	1.084×10^{12} km^3
Mass	5.9742×10^{24} kg
Density	5.52 of water
Surface gravity	9.78 m s^{-2}
Escape velocity	11.18 km s^{-1}
Day length	23.9345 hrs
Year length	365.256 days
Axial inclination	23.44°
Age	4,600 million years approx.
Distance from the Sun	Min. 147 million km
	Max. 152 million km
Surface area	5,096 million km^2
Land surface	148 million km^2
Ocean cover	71% of surface
Atmosphere	N_2 78% O_2 21%
Continental crust	35 km thick, average
Oceanic crust	7 km thick, average
Lithosphere	To 75 km depth
Mantle (silicates)	2,900 km thick
	About 3,000 °C at base
Outer core (molten iron)	2,200 km thick
	About 4,000 °C at base
Inner core (solid iron)	1,200 km thick
	Up to 5,000 °C at centre

Chapter 2
Deep time

Space is big, really big ... You may think it's a long way down the street to the chemists' but that's just peanuts to space.

Douglas Adams, *The Hitchhiker's Guide to the Galaxy*

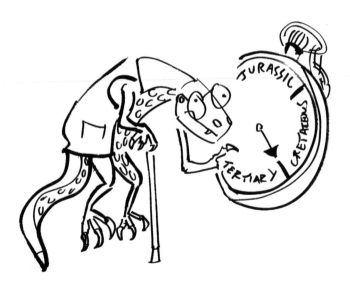

The world is not only large in its spatial dimensions. It also extends almost unimaginably far back in time. It is impossible to get a full grasp of the concepts and processes at work in geology without an understanding of what writers John McPhee, Stephen Jay Gould, and Henry Gee have referred to as deep time.

Most of us know our parents, many remember our grandparents. Only a few have met great grandparents. Their youth lies more than a century in our past, a time which seems alien to us with our vastly different scientific understanding and social structure. Just a dozen generations back, England was ruled by Queen Elizabeth I, motorized transport and electronic communication was undreamt of, and Europeans were exploring the Americas for the first time. Thirty generations takes us a thousand years back, before the Normans invaded Britain. It is also before continuous written records are likely to be able to trace our direct ancestry. We may be able to tell from archaeology and genetics roughly who our ancestors were at this time and where they might have lived but we cannot be certain. Fifty generations ago, the Roman Empire was in full swing. And 150 generations back, the Great Pyramid of Ancient Egypt had not been constructed. About 300 generations takes us back to the Neolithic in Europe at a time when the last Ice Age had only just ended and simple agriculture was the latest technological revolution. It is unlikely that archaeology can reveal where our ancestors were living at that time, though comparisons of our maternally inherited mitochondrial DNA may indicate the broad region. Add another zero to the year and we have gone back 3,000 generations to 100,000 years ago. At this time, we cannot trace separate ancestry of any living racial group. Mitochondrial DNA suggests that there was a single maternal ancestor of all modern humans in Africa not long before. But, in geological time, this is still recent.

Ten times older at a million years and we start to lose track of the modern human species. Another factor of ten and we are looking at the fossil remains of early ape ancestors. This far back it's impossible to point even to a single species and say with certainty that amongst these individuals was our ancestor. Multiply by ten again and, 100 million years ago, we are in the age of the dinosaurs. The ancestor of humans must be some insignificant shrew-like creature. A thousand million years ago and we are back amongst the first fossils, maybe before even the first recognizable animals. Ten billion years ago and we are before the birth of the Sun and solar system, at a time when the atoms that today make up our planet and ourselves were being cooked in the nuclear furnaces of other stars. Time is indeed deep.

Now think again of the changes that can take place in a few generations. Historical time is trivial compared to the age of the Earth, yet a few centuries have seen many volcanic eruptions, cataclysmic earthquakes, and devastating landslides. And think of the relentless progress of less devastating changes. In 30 generations, parts of the Himalayas have risen by maybe a metre or more. But at the same time they have eroded, probably by more than this. Islands have been born, others washed away. Some coasts have eroded back hundreds of metres, others have been left high and dry. The Atlantic has widened by about 30 metres. Now multiply all these comparatively recent changes by factors of ten or a hundred or a thousand, and you are beginning to see what can happen over geological 'deep time'.

Flood and uniformity

Humans have noticed fossil remains since prehistoric times. There are ancient stone tools that appear to have been chipped so as to show off a fossil shell. The fossilized stem of a giant cycad was placed in an ancient Etruscan burial chamber. But attempts to understand the nature of fossils are comparatively recent. The science of geology arose primarily in Christian Europe where beliefs based on biblical stories made it unsurprising to discover the shells and bones of extinct creatures high up in mountainous regions: they were the remains of animals that perished in the biblical flood. Even granite, it was suggested by so-called neptunists, was precipitated from an ancient ocean. The idea of extreme acts of God such as the flood helped people to imagine that the Earth had been shaped by catastrophes, and this was the generally accepted theory until the end of the 18th century.

In 1795 the Scottish geologist James Hutton published his now famous *Theory of the Earth*. The much quoted though paraphrased summary of its message is that 'the present is the key to the past'. This is the theory of gradualism or uniformitarianism, which says that if you want to understand geological processes you must look at the almost imperceptibly slow changes occurring today and then simply trace them out through history. It was a theory developed and championed by Charles Lyell, who was born in 1797, the year Hutton died. Both Hutton and Lyell tried to put religious beliefs in events such as the creation and the flood to one side and proposed that the gradual processes at work on the Earth were without beginning or end.

Dating creation

Attempts to calculate the age of the Earth came originally out of theology. It is only comparatively recently that so-called creationists have interpreted the Bible literally and therefore believe that Creation took just seven 24-hour days. St Augustine had argued in his commentary on Genesis that God's vision is outside time and therefore that each of the days of Creation referred to in the Bible could have lasted a lot longer than 24 hours. Even the much quoted estimate in the 17th century by Irish Archbishop Ussher that the Earth was created in 4004 BC was only intended as a minimum age and was based on carefully researched historical records, notably of the generations of patriarchs and prophets referred to in the Bible.

The first serious attempt to estimate the age of the Earth on geological grounds was made in 1860 by John Phillips. He estimated current rates of sedimentation and the cumulative thickness of all known strata and came up with an age of nearly

96 million years. William Thompson, later Lord Kelvin, followed this with an estimate based on the time it would have taken the Earth to cool from an originally hot molten sphere. Remarkably, the first age he came up with was also very similar at 98 million years, though he later refined it downwards to 40. But such dates were considered too recent by uniformitarianists and by Charles Darwin, whose theory of evolution by natural selection required more time for the origin of species.

By the dawn of the 20th century, it had been realized that additional heat might come from radioactivity inside the Earth and so geological history, based on Kelvin's idea, could be extended. In the end, however, it was an understanding of radioactivity that led to the increasingly accurate estimates of the age of the Earth that we have today. Many elements exist in different forms, or isotopes, some of which are radioactive. Each radioactive isotope has a characteristic half-life, a time over which half of any given sample of the isotope will have decayed. By itself, that's not much use unless you know the precise number of atoms you start with. But, by measuring the ratios of different isotopes and their products it is possible to get surprisingly accurate dates. Early in the 20th century, Ernest Rutherford caused a sensation by announcing that a particular sample of a radioactive mineral called pitchblende was 700 million years old, far older than many people thought the Earth to be at that time. Later, Cambridge physicist R. J. Strutt showed, from the accumulation of helium gas from the decay of thorium, that a mineral sample from Ceylon (now Sri Lanka) was more than 2,400 million years old.

Uranium is a useful element for radio dating. It occurs naturally as two isotopes – forms of the same element that differ only in their number of neutrons and hence atomic weight. Uranium-238 decays via various intermediaries into lead-206 with a half-life of 4,510 million years, whilst uranium-235 decays to lead-207 with a 713-million-year lifetime. Analysis of the ratios of all four in rocks, together with the accumulation of helium that comes from the

Some radioisotopes used for dating

Isotope	Product	Half-life	Use
Carbon 14	Carbon 12	5,730 years	Dating organic remains up to 50,000 years
Uranium 235	Lead 207	704 million years	Dating intrusions and individual mineral grains
Uranium 238	Lead 206	4,469 million years	Dating individual mineral grains in ancient crust
Thorium 232	Lead 208	14,010 million years	As above
Potassium 40	Argon 40	11,930 million years	Dating volcanic rocks
Rubidium 87	Strontium 87	48,800 million years	Dating granitic igneous and metamorphic rocks
Samarium 147	Neodymium 143	106,000 million years	Dating basaltic rocks and very ancient meteorites

decay process, can give quite accurate ages and was used in 1913 by Arthur Holmes to produce the first good estimate of the ages of the geological periods of the past 600 million years.

The success of radio-dating techniques is due in no small way to the power of the mass spectrometer, an instrument which can virtually sort individual atoms by weight and so give isotope ratios on trace constituents in very small samples. But it is only as good as the assumptions that are made about the half-life, the original abundances of isotopes, and the possible subsequent escape of decay products. The half-life of uranium isotopes makes them good for dating the earliest rocks on Earth. Carbon 14 has a half-life of a mere 5,730 years. In the atmosphere it is constantly replenished by the action of cosmic rays. Once the carbon is taken up by plants and the plants die, the isotope is no longer replenished and the clock starts ticking as the carbon 14 decays. So it is very good for dating wood from archaeological sites, for example. However, it turns out that the amount of carbon 14 in the atmosphere has varied along with cosmic ray activity. It is only because it has been possible to build up an independent chronology by counting the annual growth rings in trees that this came to light and corrections to carbon dating of up to 2,000 years could be made.

The geological column

Look at a section of sedimentary rocks in, for example, a cliff face and you will see that it is made up of layers. Sometimes annual layers corresponding to floods and droughts are visible. More often, the layers represent occasional catastrophic events or slow but steady sedimentation across hundreds of thousands or even millions of years, followed by a change of environment leading to a layer of slightly different rock. In the case of a really deep section of ancient rock, such as that seen in the Grand Canyon in Arizona, hundreds of millions of years of deposits are represented. It is a natural human instinct to divide up and classify things, and sedimentary rock with its many layers is an obvious candidate. But,

Main divisions of geological time

Eon	Era	Period			Epoch	
Phanerozoic	Cenozoic	Quaternary			Holocene	0.01
					Pleistocene	1.8
		Neogene			Pliocene	5.3
				Tertiary	Miocene	23.8
		Palaeogene			Oligocene	33.7
					Eocene	54.8
					Palaeocene	65.0
	Mesozoic	Cretaceous			UPPER	
					LOWER	142
		Jurassic			UPPER	
					MIDDLE	
					LOWER	205.7
		Triassic			UPPER	
					MIDDLE	
					LOWER	248.2
	Palaeozoic	Permian			UPPER	
					LOWER	290
		Carboniferous	Pennsylvanian		UPPER	323
			Mississippian		LOWER	354
		Devonian			UPPER	
					MIDDLE	
					LOWER	417
		Silurian			UPPER	
					LOWER	443
		Ordovician			UPPER	
					MIDDLE	
					LOWER	495
		Cambrian			UPPER	
					MIDDLE	
					LOWER	545
Precambrian	Proterozoic					2500
	Archaean					4000
	Hadean					4560

5. The main divisions of geological time (not to scale). Ages (on the right, in millions of years before present) are those agreed by the International Commission on Stratigraphy in 2000.

when viewing a spatially narrow cliff face of flat layers, it's easy to forget that the layers are not continuous around the world. The entire globe was never covered by a single shallow ocean depositing similar sediments! Just as today, there are rivers, lakes, and seas, deserts, forests, and grasslands, so in ancient times there was a panoply of sedimentary environments.

It was an English civil engineer, William Smith, who, in the early 19th century, began to make sense of it all. He was surveying for Britain's new canal network and started to realize that rocks in different parts of the country sometimes contained similar fossils. In some cases the rock types too were the same, sometimes only the fossils were similar. This enabled him to correlate the rocks in different places and work out an overall sequence. As a result, he published the first geological map. Once the dates were added in the 20th century, and the rocks correlated between different continents, it was possible to publish a single sequence of layers representing periods of geological time for the whole world. The geological column we know today is the product of many techniques, refined over the years and agreed by international collaboration.

Extinctions, unconformities, and catastrophes

It became clear that some of the changes in the geological column were bigger than others, and these provided convenient places to divide the geological past into separate eras, periods, and epochs. Sometimes there was a sudden and significant change in the nature of the rocks across such a boundary, indicating a major environmental change. Sometimes there was what is known as an unconformity, a break in deposition, caused, for example, by a change in sea level so that either deposition stopped or the layers were eroded away before the column continued. They are often also marked by major changes in fauna, represented by fossils, with many species becoming extinct and new ones beginning to arise.

A few intervals in the geological record stand out for the severity of the extinctions across them. The end of the Cambrian period and the end of the Permian period were both marked by the extinctions of around 50% of families and up to 95% of individual species of marine invertebrates. The extinctions that marked the late Triassic and late Devonian saw the loss of about 30% of families and, slightly smaller at 26%, but the most recent and the most famous, is the mass extinction at the end of the Cretaceous period 65 million years ago. That so-called K/T boundary is famous not only because it saw the extinction of the last of the dinosaurs but also because there is good evidence for the cause.

Threat from space

The first suggestion, by Walter and Louis Alvarez, that the extinction might be due to an astronomical impact at first received little scientific support. However, they soon discovered that sediments in a narrow band at that point in the geological column were enriched in iridium, an element abundant in some types of meteorite. But there was no sign of an impact crater of that age. Then evidence began to emerge, not from the land but from the sea just off the Yucatan Peninsula of Mexico, of a buried crater 200 kilometres across. There is evidence of debris from a much wider area. If, as is calculated, it marks the point where an asteroid or comet, maybe 16 kilometres across, hit the Earth, the results would indeed have been devastating. Apart from the effects of the impact itself and the tsunami that resulted, so much rock would have been vaporized that it would have spread round the Earth in the atmosphere. At first it would have been so hot that its radiant heat would have triggered forest fires on the ground. The dust would have stayed in the atmosphere for several years, blocking out sunlight, creating a global winter, and causing food plants and plankton to die. The sea bed at the impact site included rocks rich in sulphate minerals and these would have vaporized, leading to a deadly acid rain when it washed out of the atmosphere again. It is almost surprising that any living creatures survived.

The menace within

It was once hard to understand how any mass extinctions could have occurred. Now, there are so many competing theories that it is difficult to choose between them. They mostly involve severe climate change, whether triggered by a cosmic impact, changing sea levels, ocean currents and greenhouse gases, or a cause from within the planet such as rifting or major vulcanism. It does seem that most of the mass extinctions we know coincided at least approximately with major eruptions of flood basalts. In the case of the late Cretaceous, it was the eruptions that produced the Deccan Traps in western India. There has even been a suggestion that a major asteroid impact caused shock waves to focus on the other side of the Earth, triggering eruptions. But the times and positions do not seem to line up well enough to prove that explanation. Whatever the reason, the history of life and of the planet has been punctuated by some catastrophic events.

Chaos reigns

We can all remember climatic events that stand out, say over the last decade, as the worst winter, flood, storm, or drought. Take the record back for a century and the likelihood is that an even bigger one will stand out. Authorities often use the concept of a '100-year' flood in planning coastal or river flood defences; they are designed to withstand the sort of flood that only happens once a century. It's likely to be more severe than the sort which happens only once a decade. But, if you extend the same idea to a thousand years or a million years, there is likely to be one that will be bigger still. According to some theorists, that is likely to be true of anything from floods, storms, and droughts to earthquakes, volcanic eruptions, and asteroid impacts. Over geological time we had better watch out!

Deeper time

The list of geological periods that is often shown in books goes back
only about 600 million years to the start of the Cambrian period.
But that ignores 4 billion years of our planet's history. The trouble
with most Pre-Cambrian rocks is that they are, as Professor Bill
Schopf of the University of California puts it, fubaritic – fouled up
beyond all recognition. The constant tectonic reprocessing of the
Earth from within, and the relentless pounding of weather and
erosion from above, mean that most of the Pre-Cambrian rocks that
survive at all are deeply folded and metamorphosed. But on most
clear nights you can see rocks that are more than 4 billion years
old – by looking up at the Moon rather than down at the Earth. The
Moon is a cold, dead world with no volcanoes and earthquakes,
water or weather to resurface it. Its surface is covered with impact
craters, but most of those happened early in its history when the
solar system was still full of flying debris.

The Pre-Cambrian rocks that do survive on Earth tell a long and
fascinating story. They are not, as Darwin had supposed, devoid of
the traces of life. Indeed, the end of the Pre-Cambrian, from about
650 to 544 million years ago, has yielded a rich array of strange
fossils, particularly from localities in southern Australia, Namibia,
and Russia. Prior to that there seems to have been a particularly
severe period of glaciation. The phrase 'snowball Earth' has been
used, conveying the possibility that all the world's oceans froze over.
Inevitably, that would have been a major setback for life, and there
is scant evidence for multicellular life forms before this. But there is
abundant evidence for microorganisms – bacteria, cyanobacteria,
and filamentous algae. There are filamentous microfossils from
Australia and South Africa that are around 3,500 million years old,
and there is what looks like the chemical signature of life in carbon
isotopes in rocks from Greenland that are 3,800 million years old.

During the first 700 million years of its history, the Earth must have
been particularly inhospitable. There were numerous major

impacts far bigger than that which may have killed the dinosaurs. The scars of this late heavy bombardment can still be seen in the great Maria basins on the Moon, which are themselves giant impact craters filled with basalt lava melted by the impacts. Such impacts would have melted much of the Earth's surface and certainly vaporized any primitive oceans. It is possible that the water on our planet today came from a subsequent rain of comets as well as from volcanic gases.

Dawn of life

The early atmosphere of Earth was once thought to have been a mixture of gases such as methane, ammonia, water, and hydrogen, a potential source of carbon to primitive life forms. But it is now believed that strong ultraviolet radiation from the young Sun would have broken that down quickly to give an atmosphere of carbon dioxide and nitrogen. No one yet knows for certain how life began. There are even claims that it may have had an extra-terrestrial origin, arriving on Earth in meteorites from Mars or beyond. But laboratory studies are beginning to show how some chemical systems can begin to self-organize and catalyse their own reproduction. Analysis of present-day life forms suggests that the most primitive are not the sort of bacteria that scavenge organic carbon or that use sunlight to help them photosynthesize but those that use chemical energy of the sort that is found today in deep-sea hydrothermal vents.

By 3,500 million years ago, there were almost certainly microscopic cyanobacteria and arguably primitive algae – the sort of thing we see today in pond scum. These began to have a dramatic effect. Using sunlight to power photosynthesis, they took in carbon dioxide from the atmosphere, effectively eating the blanket that, by the greenhouse effect, kept the Earth warm when the power of the Sun was weak. This may be what led ultimately to the late Pre-Cambrian glaciation. But long before that, it resulted in the worst pollution incident the world has known. Photosynthesis released a gas that had not existed on Earth before and which was probably toxic to many life forms: oxygen. At first, it did not last long in the atmosphere but quickly reacted with dissolved iron in sea water, resulting in thick deposits of banded iron oxide. Almost literally, the world went rusty. But photosynthesis continued, and free oxygen began to build up in the atmosphere from about 2,400 million years ago, paving the way for animal life that could breathe the oxygen and eat the plants.

Birth of Earth

About 4,500 million years ago there was a great cloud of gas and dust, the product of several previous generations of stars. It began to contract under gravity, perhaps boosted by the shock waves from a nearby exploding star or supernova. A slight rotation in the cloud accelerated as it contracted and spread the dust out into a flattened disc around the proto-star. Eventually, the central mass, mostly of hydrogen and helium, contracted sufficiently to trigger nuclear fusion reactions at its core and the Sun began to shine. A wind of charged particles began to blow outwards, clearing some of the surrounding dust. In the inner part of the nebula, or disc, only refectory silicates remained. Further out, the hydrogen and helium accumulated to form the giant gas planets Saturn and Jupiter. Volatile ices such as water, methane, and nitrogen were driven still further out and formed the outer planets, Kuiper belt objects, and comets.

The inner planets – Mercury, Venus, the Earth, and Mars – formed by a process known as accretion which began as particles bumped into one another, sometimes splitting, occasionally joining together. Eventually, the larger lumps developed sufficient gravitational attraction to pull others to them. As the mass increased, so did the energy of the impacts, melting the rocks so that they began to separate out, with the densest, iron-rich minerals sinking to form a core. The new Earth was hot, probably at least partially molten, from the impacts, from the energy released by its gravitational contraction, and from the decay of radioactive isotopes. It is likely that many radioactive elements in the pre-solar nebula had been created not long before in supernovae explosions and would still have been radioactively hot. So it is hard to see how there could have been liquid water on the surface initially, and it is possible that the first atmosphere was mostly stripped away by the force of the solar wind.

A chip off the block

The formation of the Moon had long been a mystery to science. Its composition, orbit, and rotation didn't fit with the idea that it had split off from the young Earth, formed alongside it, or been captured whilst passing it. But one theory does now make sense and has been convincingly simulated in computer models. It involves a proto-planet about the size of Mars crashing into the Earth about 50 million years after the formation of the solar system. The core of this projectile would have merged with that of the Earth, the force of the impact melting most of the Earth's interior. Much of the outer layers of the impactor, together with some terrestrial material, would have vaporized and been flung into space. A lot of that collected in orbit and accreted to form the Moon. This cataclysmic event gave us a companion which seems to have a stabilizing effect on the Earth, preventing its rotation axis swinging chaotically and thus making our planet a more amenable home to life.

Chapter 3
Deep Earth

The surface of the Earth is covered by a relatively thin, cold, hard crust. Beneath the oceans it is about 7 or 8 kilometres thick; in the continents, 30 to 60 kilometres thick. At its base lies the Mohorovicic discontinuity or Moho, a layer which reflects seismic waves, probably as a result of a change in composition to the dense rocks of the mantle beneath. The lithosphere, the complete slab of cold, hard material on the Earth's surface, includes not only the crust but the top of the mantle as well. In total, the continental lithosphere may be 250 or even 300 kilometres thick. It thins under the oceans, as you approach the mid-ocean ridge system, down to little more than the 7-kilometre crust. The lithosphere is not a single rigid layer, however. It is split into a series of slabs called tectonic plates. They are our principal clue as to how the deep Earth works. To understand what's going on, we must probe beneath the crust.

Digging deep

Only 30 kilometres away from us lies a place we can never visit. If the distance was horizontal it would just be an easy bus ride away, but the same distance beneath our feet is a place of almost unimaginable heat and pressure. No mine can tunnel that deep. A proposal in the 1960s to use ocean-drilling techniques from the oil exploration industry to drill right through the ocean crust into the mantle, the so-called Moho project, was ruled out on grounds of cost and difficulty. Attempts at deep drilling on land on Russia's Kola Peninsula and in Germany had to be abandoned after about 11,000 metres. Not only was the rock difficult to drill, but the heat and pressure tended to soften the drill components and squeeze the hole shut again as soon as it was drilled.

Messengers from the deep

There is one way in which we can sample the mantle directly: in the outpourings of deep-rooted volcanoes. Most of the magma that erupts from volcanoes comes from only partial melting of the source material, so basalt, for example, is not a complete sample of mantle rock. It does, however, carry isotopic clues to what lies beneath. For example, basalt from some deep-rooted volcanoes, such as that in Hawaii, contains helium gas with a high ratio of helium 3 to helium 4, as the early solar system is believed to have had. So this is thought to come from a part of the Earth's interior that is still pristine. The helium gets lost in volcanic eruptions and is slowly replaced by helium 4 from radioactive decay. The basalt in ocean ridge volcanoes is depleted in helium 3. This suggests that it is recycled material that lost helium gas in earlier eruptions and does not come from so deep in the mantle.

Violent volcanic eruptions do sometimes carry in their magma more direct samples of mantle rocks. These so-called xenoliths are samples of mantle rock that have not been melted, just carried along in the flow. They are typically dark, dense, greenish rocks such as peridotite, rich in the mineral olivine, a magnesium/iron silicate. Similar rock is sometimes found in the deep cores of mountain ranges which have been thrust up from great depths.

Slow flow

The magnificent medieval stained-glass windows of Canterbury Cathedral can tell us something about the nature of the Earth's mantle. The windows are composed of many small panes of coloured glass in a leaded frame. If you look at the sunlight filtered through the panes, you will notice that some of them are darker at the bottom than at the top. This is because the glass flows. Technically, it's a super-cooled fluid. Over the centuries, gravity has made the panes slowly sag so that the glass is thicker at the bottom. Yet, to the touch, or, heaven forbid, a hammer blow, the glass still behaves as a solid. A key to understanding the Earth's mantle is the realization that the silicate rocks there can flow in the same sort of way, even though they do not melt. In fact, the individual mineral grains are constantly re-forming, giving rise to the motion known as creep. The effect is that the mantle is very, very viscous, like extremely thick, sticky treacle.

A planetary body scan

The clearest clues to the internal structure of the Earth come from seismology. Earthquakes send out seismic shock waves through the planet. Like light being refracted by a lens or reflected by mirrors, seismic waves travel through the Earth and reflect off different layers within it. Seismic waves travel at different velocities depending on how hot or soft the rock is. The hotter and therefore softer the rock, the slower the wave travels. There are two main types of seismic wave, primary, or P, waves, which are the faster and thus the first to arrive at a seismograph, and secondary, or S, waves. P waves are pressure waves with a push-pull motion; S waves are shear waves and cannot travel through liquids. It was by studying S waves that the molten outer core of the Earth was first revealed. Detecting these seismic waves on a single instrument would not tell you much, but today there are networks of hundreds of sensitive seismometers, strung out around the planet. And every day there are many small earthquakes to generate signals. The result is a bit like a body scanner in a hospital, in which the patient is surrounded by X-ray sources and sensors and computers use the results to build up a 3D image of her internal organs. The hospital version is known as a CAT scan, standing for Computer Assisted Tomography. Its whole-Earth equivalent is called seismic tomography.

The global network of seismographs is best at seeing things on a global scale. It will reveal the overall layering in the mantle and changes in seismic velocities due to high or low temperature on scales of hundreds of kilometres. There are also more closely spaced arrays, originally set up to detect underground nuclear tests, and they, together with new arrays being deployed by geophysicists in geologically interesting regions, have the potential to see structure deep in the mantle on a scale of a few kilometres. And it seems that there is structure on every scale. The clearest things in these whole-Earth body scans are the layers. Below 2,890 kilometres, the depth of the liquid outer core of our planet, S waves will not pass. But several features stand out within the mantle. There is, as we've mentioned, the Mohorovicic discontinuity at the base of the crust, and another at the base of the hard lithosphere. The asthenosphere beneath is softer so the seismic velocities are slower. There are clear layers 410 kilometres down and 660 kilometres down, with another less distinct layer after about 520 kilometres. At the base of the mantle is another, probably discontinuous, layer called the

D″ or D double prime layer, which varies from nothing to about 250 kilometres thick.

Seismic tomography also reveals more subtle features. Essentially, colder rock is harder so seismic waves travel through it more quickly than they do through hotter, softer rock. Where old, cold ocean crust dives beneath a continent or into an ocean trench, reflections from the descending slab reveal its passage down into the mantle. Where Earth's hot core bakes the underside of the mantle, it appears to soften and rise in a huge plume.

The mantle is full of mysteries which at first sight seem to be contradictory. It is solid yet it can flow. It's made up of silicate rock which is a good insulator, yet somehow about 44 terawatts of heat finds its way to the surface. It's hard to see how that heat flow could happen through conduction alone, and yet, if there was convection, the mantle would be mixed, so how could it show a layered structure? And how could ocean volcanoes erupt magma with a different mix of tracer isotopes than that believed to exist in the bulk of the mantle, unless there are unmixed regions or layers? Resolving these mysteries has been one of the prime areas of geophysics in recent years.

A diamond window on the mantle

Some of the best clues have come from understanding the nature of the rocks down there. To find what the rocks are like deep within the Earth, you have to replicate the fantastic pressures down there. Amazingly, that's possible with just your finger and thumb. The trick is to get hold of two good, gem-quality diamonds, cut in what jewellers term 'brilliant' cuts, with a tiny, perfectly flat face at the apex of each. Mount them face to face with a microscopic rock sample between the two, and turn a little thumbscrew to force the faces tighter together. The force gets so concentrated between the tiny diamond anvils that it's possible to create pressures more than 3 million times atmospheric pressure (300 gigapascals), just by turning the screw. Because the diamonds are conveniently transparent, the sample can be heated by shining a laser in, and viewed with a microscope and other instruments. This can literally be a window on what rocks are like deep in the mantle.

Professor Bill Bassett was studying a tiny crystal in a diamond anvil one day in his lab at Cornell University. Nothing much had happened when he'd increased the pressure, so he decided to go for lunch. As he was leaving, he heard a sudden 'crack' from the anvil. Certain that one of his precious diamonds had broken, he rushed back and looked down the microscope. The gems were OK, but the sample had suddenly transformed into a new, high-pressure crystal form. It was what is known as a phase change: the composition remains the same but the structure changes, in this case into a more dense crystal lattice.

We know from the composition of xenoliths that at least the upper mantle is made of rocks such as peridotite, rich in the magnesium and iron silicate mineral olivine. Put a tiny sample of this between diamond anvils and turn up the pressure and it goes through a whole series of phase changes. At a pressure of about 14 gigapascals, equivalent to a depth in the mantle of about 410 kilometres, olivine transforms into a new structure called wadsleyite. At 18 gigapascals, 520 kilometres down, it changes again, adopting the structure of ringwoodite, a form of the mineral spinel. That then changes at 23 gigapascals, corresponding to 660 kilometres down, into two minerals, perovskite and a magnesium iron oxide mineral called magnesiowüstite. You'll notice that the phase changes happen at precisely the depths at which seismic waves can be reflected. So perhaps the layers indicate a change in crystal structure rather than composition.

A double boiler?

The 660-kilometre layer, the division between the upper and lower mantle, is a particularly strong feature and the focus of vigorous debate between those who think that the entire mantle is circulating in a huge convection system and those who think that it is more like a double boiler with separate circulating cells in the upper and lower mantle and little or no exchange of material between them. Historically, geochemists tend to favour the double structure as it allows for chemical differences between the layers, whereas geophysicists prefer whole-mantle convection. Present indications are that both might be right, in a compromise solution in which whole-mantle circulation is possible but difficult. Data from seismic tomography would seem at first to favour the double boiler idea. The seismic scans reveal where slabs of subducted ocean crust sink down towards the 660 km anomaly. But they do not seem to pass through it. Rather, the material spreads out and seems to collect at that depth, for hundreds of millions of years. But further scans show where it can break through like an avalanche and continue on through the lower mantle almost to the top of the core.

In June 1994, Bolivia was shaken by a powerful earthquake. It did little damage because its focus was so deep – about 640 kilometres. But at that depth, the rocks should be too soft to fracture. This is a region where a slab of old ocean crust from the Pacific is sinking down beneath the Andes. What must have happened is that a whole layer of rock underwent a catastrophic phase change into the denser perovskite structure. That seems to be necessary before it can sink down into the lower mantle. The explanation solves the mysteries of mantle layering and deep earthquakes at one go.

But there is much that still needs explaining. For example, the slab of ocean crust that is subducting below the Tonga trench in the Pacific is passing into the mantle at about 250 millimetres per year, far too fast for its temperature to even out. Material would reach the base of the upper mantle in just 3 million years and its low temperature should be obvious if it pools there or extends into the lower mantle. But there is no evidence for such a slab. One theory is that not all of the olivine converts into higher-density minerals,

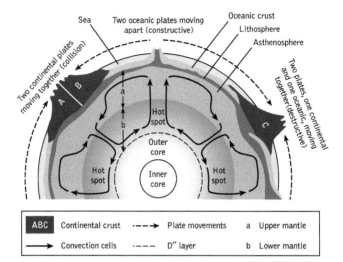

6. The basic circulation in the Earth's mantle and how it is reflected in lithospheric plate motions and plate boundaries. For clarity, motions are simplified and the vertical scale of the lithosphere is greatly exaggerated.

making the old slab neutrally buoyant in the upper mantle. The combination of cool temperature and mineral content would give it a seismic velocity very similar to other mantle material, so it would not show up easily, just as a layer of glycerine does not show up well in water. There is indeed tantalizing faint seismic evidence for such a slab deep below Fiji.

Message in a diamond

Diamond is the high-pressure form of carbon. It can only form in the Earth at depths of over 100 kilometres, sometimes well over this. Isotope ratios in diamonds suggest that they often form from carbon in subducted ocean crust, maybe carbonate from ocean sediments. Sometimes, there are tiny inclusions of other minerals within a diamond. It is not a feature that is popular among gem stone dealers, but it is just what geochemists are searching for.

Minute analysis of those inclusions can tell the long and sometimes tortuous history of the diamond's formation and passage through the mantle.

Some of the inclusions are of a mineral called enstatite, which is a form of magnesium silicate. Some researchers believe that it was originally magnesium silicate perovskite and comes from the lower mantle. Their evidence comes from the observation that it contains only one-tenth as much nickel as would be expected in the upper mantle. At lower mantle temperatures and pressures, nickel gets taken up into a mineral called ferropericlase, which is also a common inclusion of diamonds, leaving very little nickel left in magnesium silicate perovskite. In a few cases, the inclusions are rich in aluminium which, under upper mantle conditions, is locked up in garnet. And some inclusions are iron-rich, suggesting that they might have originated very deep in the mantle, close to the core mantle boundary. These deep diamonds also have a different carbon isotope signature, believed to be characteristic of deep mantle rock rather than subducted ocean lithosphere. Estimates of the age of diamonds and the rock that surrounds them suggests that some have had a very long and tortuous passage through the mantle that may have taken them more than a billion years. But it is convincing evidence of at least some transfer between the lower and upper mantle.

Almost as fascinating as the diamonds themselves is the rock in which they are found. It's called kimberlite after the South African diamond-mining town of Kimberley. The rock itself is a mess! Apart from the diamonds, it contains a whole range of angular lumps and pulverized fragments of different rocks; a so-called breccia. It is volcanic and tends to form the carrot-shaped plugs of ancient volcanic vents. It is hard to determine its exact composition because it contains so much pulverized debris from its passage through the lithosphere, but the original magma must have been mostly olivine from the mantle together with an unusual amount of volatile material now in the form of mica. If it had found its way up slowly from the mantle, we would not have diamonds today. Diamond is unstable at pressures found less than 100 kilometres underground and, given time, would dissolve in the magma. But kimberlite volcanoes did not keep it waiting. It is estimated that the average speed of material through the lithosphere was about 70 kilometres per hour. The widening neck of the vent as it approaches the surface suggests that volatile material was expanding explosively and the surface eruption speed could have been supersonic. As a result, all the rock fragments collected on the way up have been quenched, frozen in time, so that they represent samples from deep in the lithosphere and even the mantle.

The base of the mantle

Recent analysis of seismic data from around the world has revealed a thin layer at the base of the mantle, the D″ layer, up to 200 kilometres thick. It is not a continuous layer but seems more like a series of slabs, a bit like continents on the underside of the mantle. This could be regions where silicate rocks in the mantle are partly mixed with iron-rich material from the core. But another explanation is that this is where ancient ocean lithosphere comes to rest. After its descent through the mantle, the slab is still cold and dense so it spreads out at the base of the mantle and is slowly heated by the core until, perhaps a billion years later, it rises again in a mantle plume to form new ocean crust.

Clues to the deep interior of the Earth also come from measuring tiny variations in day length. Our spinning planet is gradually slowing down due to the pull of the moon on the tides and to the rising of land compressed by ice in the last Ice Age. But there are other even smaller variations of a few billionths of a second. Some may be due to atmospheric circulation blowing on mountain ranges like wind on a sail. But there is another component which seems to be caused by circulation in the outer core pushing on ridges in the base of the mantle like ocean currents pushing on the keel of a ship. So there may be ridges and valleys like upside-down mountain ranges on the base of the mantle. There seems to be a great depression in the core beneath the Philippines that is 10 kilometres deep, twice the depth of the Grand Canyon. Bulging up beneath the Gulf of Alaska is a high spot on the core; a liquid mountain taller than Everest. Maybe sinking cold material indents the core, while hotspots bulge up.

Super plumes

Although much hotter, the perovskite rock of the lower mantle is much more viscous than upper mantle rocks. Estimates suggest that it is 30 times more resistant to flow. As a result, material rises from the base of the mantle in a much slower, broader column than the plumes which characterize the upper mantle. It behaves, in very, very slow motion, rather like the blobs of gloop in a lava lamp. It may well be true that, although some material circulates through the entire mantle, there are also smaller convection cells that are confined to the upper mantle. Convection cells in experimental systems tend to be about the same width as they are deep and, in some parts of the world at least, the spacing of plumes of mantle material seem to match the 660-kilometre depth of the upper mantle.

How the Earth melts

What goes down must come back up again. As plumes of hot mantle rock slowly rise towards the crust, the pressure on them drops and they begin to melt. Scientists can recreate what happens using great hydraulic presses to squeeze samples of artificial rock, heated in furnaces. It's not the entire rock that melts, only a few per cent; producing magma that is less dense than the rest of the mantle and so is able to rise up rapidly to the surface and erupt as basalt lava. How it flows through the remaining rock was another great mystery. It turns out to be down to the microscopic structure of the rock. If the angles at the corners of the little pockets of melt that form between rock grains were large, the rock would be like a Swiss cheese; the pockets would not interconnect and the melt couldn't flow out. But those angles are small and the rock is like a sponge, with all the pockets interconnecting. Squeeze the sponge and the liquid flows. Squeeze the mantle and the magma erupts.

Free-fall

When he saw an apple fall, Isaac Newton realized that the force of gravity was pulling objects towards the centre of the Earth. What he did not know was that apples fall slightly faster in some parts of the world than others – not that it is a difference you normally notice, nor could you easily measure it with apples. But you can with spacecraft. The secret of flying, according to Douglas Adams' *Hitchhiker's Guide to the Galaxy*, is to fall but forget to hit the ground. That's roughly what a satellite does. It's falling freely, but its speed keeps it in orbit. The stronger gravitational pull of a region of dense rock will make satellites speed up. Over a region of lower gravity, they will slow down. By tracking the orbits of low satellites, geologists can build up gravitational maps of the Earth beneath.

When geophysicists compared gravity maps of the surface of the Earth with seismic tomography scans of its interior, they had a surprise. You might expect that cold, dense slabs of ocean crust would result in an excessive gravitational pull because of their higher density, whilst a plume of hot mantle rock rising upwards would be less dense and cause a gravity low. That reality is the opposite way around. The effect is especially pronounced over southern Africa, where a huge plume of hot mantle appears to be rising, and around Indonesia, where cold slabs are sinking. Brad Hager of the Massachusetts Institute of Technology came up with an explanation. The mantle super-plume under southern Africa is causing a huge part of the continent to rise up, higher than you would expect were it simply floating on a static mantle. Southern Africa, he estimates, is elevated by about 1,000 metres above where it would naturally float on the mantle, and this excess uplift of rock causes the gravity high. Similarly, the subducting ocean lithosphere beneath Indonesia is dragging the surrounding surface down behind it, creating a gravity low and resulting in a general rise of sea level compared to the land. Clement Chase, now at the University of Arizona, realized that other broad gravity anomalies corresponded to the ghosts of past subduction. A long band of low gravity that passes from Hudson Bay in Canada, over the North Pole, through Siberia and India, and on to the Antarctic seems to mark a series of subduction zones where ancient sea floor has plunged back into the mantle over the last 125 million years. What was thought to be a rise in sea level which submerged most of the eastern half of Australia about 90 million years ago may have been caused by the continent drifting over an ancient subduction zone that tugged at the region as it passed over, lowering land by more than 600 metres.

The core

We have no direct experience or samples of the Earth's core. But we do know from seismic waves that the outer part of it is liquid and only the inner core is solid. We also know that the core has a much higher density than the mantle. The only material that is dense enough and sufficiently abundant in the solar system to make up the bulk of the core is iron. Although we do not have samples of the Earth's core, we do have pieces of something that's likely to be similar, in iron meteorites. Though not as common as stony meteorites, they are easier to spot. They are believed to come from large asteroids in which an iron core separated out before they were smashed by bombardment early in the history of the solar system. They are mostly made of iron metal but contain between 7% and 15% of nickel. Often, they have a structure of intergrown crystals of two alloys, one containing 5% nickel, the other about 40% nickel, in proportions that give the bulk composition.

An iron core must have formed in the Earth by gravitational separation from the silicate mantle when the new Earth was at least partially molten. As the layers separated, so-called siderophile elements such as nickel, sulphur, tungsten, platinum, and gold that are soluble in molten iron would have separated with them. Lithophile elements would have been held back by the silicate mantle. Radioactive elements such as uranium and hafnium are lithophile, whereas their decay products, or daughters, are isotopes of lead and tungsten so would have been separated out into the core at its formation. That consequently reset the radioactive clock in the mantle at the time the core formed. Estimates of the age of mantle rock put that separation at 4.5 billion years ago, about 50 to 100 million years after the ages of the oldest meteorites which seem to date from the formation of the solar system as a whole.

The inner core

The centre of the Earth is frozen. Frozen at least from the viewpoint of molten iron at the incredible pressures down there. As the planet cools, solid iron crystallizes out from the molten core. Present understanding of the electrical dynamo that generates the Earth's magnetic field requires a solid iron core, but the planet may not have had one for its entire history. There is evidence of the Earth's past magnetic field locked into rocks throughout the Phanerozoic. But most Pre-Cambrian rocks have been so altered that it is difficult to measure any original magnetism. So the only estimate of the age of the inner core comes from models of thermal evolution of the core as the Earth slowly cools. It's the same sort of calculation that Lord Kelvin performed in the late 19th century to estimate the age of the Earth from its rate of cooling. But now we know there is additional heat from radioactive decay. The latest analysis suggests that the inner core began solidifying somewhere between 2.5 and 1 billion years ago, depending on its radioactive content. That may seem a long time, but it implies that for billions of years of its early history, the Earth was without an inner core and perhaps without a magnetic field.

Today, the inner core is about 2,440 kilometres across, 1,000 kilometres smaller than the Moon. But it is still growing. The iron is crystallizing at a rate of about 800 tonnes a second. That releases a considerable amount of latent heat, which passes through the liquid outer core, contributing to the churning of the fluid within it. As the iron or iron-nickel alloy crystallizes out, impurities within the melt, mostly dissolved silicates, separate out. This material is less dense than the molten outer core, so it rises through it in a steady rain of perhaps sand-like particles. It probably accumulates on the base of the mantle like a sort of upside-down sedimentation, collecting in upside-down valleys and depressions. There are seismic hints of a very low velocity layer at the base of the mantle that this upward sedimentation could explain. The sandy sediment would trap molten iron just as ocean sediment traps water. By holding iron within it, the layer provides material that can magnetically couple the magnetic field generated in the core with the solid mantle. If some of this material rises in super-plumes to contribute to flood basalts on the surface, it could explain the high concentrations of precious metals such as gold and platinum in such rocks.

Magnetic dynamo

From the surface, the Earth's magnetic field looks as if it could be generated by a large permanent bar magnet in the core. But it is not. It must be a dynamo, with the magnetic field generated by electrical currents in the circulating molten iron of the outer core. Faraday showed that if you have an electrical conductor, any two out of electrical current, magnetic field, and motion will generate the third. That is the principle on which all electrical motors, generators, and dynamos work. But in the case of the Earth, there are no external electrical connections. Somehow both the currents and the field are generated and sustained by the convection currents in the core. This is what is called a self-sustaining dynamo. But it must have needed some sort of kick-start. Perhaps that came from the Sun's magnetic field before the Earth had one of its own.

The magnetic field on the Earth's surface is relatively simple, but the currents in the Earth's core that generate it must be far more complex. Many models have been proposed, some of which, such as

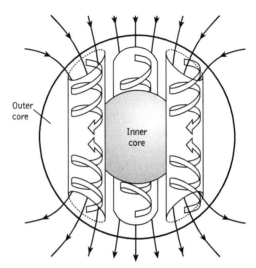

7. **A possible model for the generation of the Earth's magnetic field. Convection currents in the outer core spiral due to Coriolis forces (ribbon arrows). That, and electrical currents (not shown) produce the magnetic field lines (black arrows).**

the idea of a rotating conducting disc, are purely theoretical. A model that best accounts for the field we see involves a series of cylindrical cells each containing spiral circulation produced by the combination of thermal convection and the Coriolis forces generated by the Earth's rotation. One of the strangest features of the Earth's magnetic field, as we will see in more detail in the next chapter, is that it reverses its polarity at irregular intervals, typically of a few hundred thousand years. At other times there can be periods of up to 50 million years without a reversal. Evidence of the strength of the field trapped in individual volcanic crystals suggests that the field might have been stronger than it is today during such non-reversing periods, or superchrons. The magnetic field is not precisely aligned with the Earth's axis of rotation. At present, it is inclined at about 11 degrees to the Earth's rotation axis. But it hasn't always stayed there. In 1665 it was almost true north, then wandered off, reaching 24 degrees west by 1823. Computer models cannot explain it exactly, but suggest that the dynamo itself is fluctuating chaotically. Most of the time, coupling with the mantle damps out the effects, but sometimes they get so great that the field flips. What is not clear is whether the reversals happen virtually overnight or whether there are thousands of years in which the field wanders wildly or virtually disappears. If the latter is the case, it would be bad news not only for navigation by compass but for life on Earth generally as it would be exposed to more hazardous radiation and particles from space.

There have also been attempts to model what goes on in the Earth's core by experiment. This is not easy since it requires a large volume of electrically conducting fluid circulating with sufficient velocity to excite a magnetic field. It was achieved in Riga by German and Latvian scientists who used 2 cubic metres of molten sodium contained in concentric cylinders. By propelling the sodium down the central cylinder at 15 metres per second, they were able to generate a self-exciting magnetic field.

Taking the Earth's temperature

The deeper you go, the hotter it gets, but how hot is the middle of the Earth? The answer is that, at the boundary between the molten outer core and the solid inner core of the Earth, the temperature must be at exactly the melting point of iron. But the melting point of iron under those incredible pressures will be very different from its value on the surface of the Earth. To find out what it is, scientists must recreate those conditions in their laboratories or calculate them from theory. They've tried two different practical methods: one using tiny samples squeezed between diamond anvils, the other using a giant multi-stage compressed gas gun to compress samples just for an instant. Because of the difficulties in achieving such incredible pressures – 330 gigapascals at the inner core boundary – and because of the difficulty in calibrating pressure so that you know when you've got there, both methods have yet to measure that temperature directly. What they can do is measure the melting point of iron at slightly lower pressures and try to extrapolate downwards. But there are still difficulties. Not least because the core is not pure iron and impurities can affect the melting point. Theoretical calculations put the inner core boundary at about 6,500 degrees Celsius for pure iron, maybe 5,100–5,500 degrees Celsius for iron with the probable range of impurities in the core. These are between estimates from diamond anvil and gas gun experiments.

The study of seismic waves passing through the inner core has produced one more surprise. The waves seem to travel 3–4% faster through the inner core when going north to south compared to east to west; the inner core exhibits anisotropy, a structure or grain which is not the same in all directions. The explanation could be that the inner core is made up of lots of aligned crystals of iron – or even one big crystal more than 2,000 kilometres across! It's also possible that there could be convection within the inner core just as there is in the mantle. And there may be a small amount of liquid caught up in the crystalline mush. It has been calculated that between 3% and 10% by volume of flat discs of liquid aligned with the equator would give the inner core the anisotropy observed.

Spinning core

Like the Earth as a whole, the inner core is rotating, but not exactly in the same way as the rest of the Earth. It is in fact rotating slightly faster than the remainder of the planet, gaining nearly one-tenth of a turn in the past 30 years. Careful study of seismic waves from earthquakes in the South Sandwich Islands off the southern tip of South America that were detected in Alaska show the effect. It is revealed due to the north–south anisotropy in the inner core that we just discussed. As the inner core pulls ahead of the rest of the Earth, the effect due to that anisotropy changes. Seismic waves that skimmed past just outside the inner core arrived in Alaska just as quickly in 1995 as they did in 1967. But waves passing through the inner core made the trip 0.3 seconds faster in 1995 than in 1967, showing that the fast track axis of the inner core has been swinging into alignment at about 1.1 degrees per year. Understanding why the inner core is spinning so fast may give insight into what is going on in that strongly magnetic environment. It could be that currents in the outer core, analogous to the jet streams in the atmosphere, are putting a magnetic tug on the inner core.

So far, only about 4% of the total core has frozen. But, in 3 or 4 billion years' time, the entire core will have solidified and we may lose our magnetic protection.

Chapter 4
Under the sea

Hidden world

Water covers 71% of our planet. Only 1% of that is fresh water; 2% is frozen, and the remaining 97% is salt water in the oceans. It averages more than 4,000 metres deep and reaches down to a maximum of 11,000 metres. All we can easily see is the surface. Scarcely any sunlight penetrates deeper than 50 metres, the so-called photic zone. The rest is a cold, dark world that is alien to us, or at least it was until about 130 years ago.

In 1872, HMS *Challenger* set sail on the first scientific voyage of oceanographic exploration. She visited every ocean and travelled 100,000 kilometres in four years, but depths could only be taken up by single point soundings with a weight lowered over the side of the boat. So the pace of oceanography was slow until the development of techniques such as sonar and sediment coring during the Second World War. During the Cold War, Western powers needed good maps of the sea floor so they could conceal their own submarines, and required advanced sonar and arrays of hydrophones to detect Soviet submarines. Today, ship-mounted and towed sonar scanners have mapped much of the sea floor in considerable detail. The ocean drilling programme has sampled the underlying rocks in many areas, and deep-water manned and robot submersibles have visited many of the most interesting places. But there is still a wealth of exploration ahead.

Where did the water come from?

It seems likely that the Earth's most primitive atmosphere was largely stripped away by the strength of the newborn Sun's solar wind. It is almost certain that the heat generated by the heavy bombardments that completed the formation of the Earth and the huge impact that created the Moon must have melted the surface rocks and driven off most of the original water. So where did our vast oceans come from? There are clues in the oldest rocks, from 4 billion years ago, that liquid water was around when they formed, and there is evidence from not long after that of aquatic bacteria. The oldest fossilized imprints of rain drops are in sediments about 3 billion years old in India. Some of the Earth's surface water may have escaped from the planet's interior in volcanic gases, but most of it probably fell from space. Even today about 30,000 tons of water falls to Earth each year in a fine rain of cometary particles from deep space. In the early history of the solar system that flux must have been significantly higher, and many of the late impacts are likely to have been from whole or fragmentary comets, the composition of which has been likened to dirty snowballs, containing abundant water ice.

Salty seas

Today, about 2.9% by weight of sea water is made up of dissolved salts, mostly common salt, sodium chloride, but also sulphates and bicarbonates and chlorides of magnesium, potassium, and calcium, plus trace elements. The salinity varies depending on the evaporation rate and the influx of fresh water. So, for example, in the Baltic the salinity is low but in the landlocked Dead Sea, it is about six times the average of 35 grams of solids per kilogram of sea water. But the relative proportions of each of the main components of salinity remain constant worldwide.

The oceans were not always that salty. Most of the salt is believed to have come from rocks on land. Some of it was simply dissolved by rain and rivers and some was released by chemical weathering, in which carbon dioxide dissolves in rain to make weak carbonic acid. This slowly converts silicate minerals in rock into clay minerals. These tend to retain potassium but release sodium, which is why sodium chloride is the biggest component of sea salt. For the last few hundred million years, ocean salinity has been approximately constant, with the input of salts from weathering balanced by their deposition in evaporite deposits and other sediments.

The living ocean

The oceans also contain many chemicals in trace quantities and many of these are nutrients important for life and hence for ocean productivity. As a result, they are often depleted in surface waters. Colour scanners flown in space and tuned to the characteristic wavelengths of pigments such as chlorophyll in phytoplankton can map out the seasonal productive zones in the ocean. The highest productivity tends to occur in the spring in mid-to-high latitudes where warm water meets cold but nutrient-rich waters. In the 1980s, the late John Martin of Moss Landing Marine Laboratory in California noticed that blooms of plankton can arise down-current off volcanic ocean islands. He suggested that iron might be a limiting nutrient in ocean productivity and that the volcanic rock was supplying traces of dissolved iron. That has since been confirmed by experiments seeding patches of iron salts in the South Pacific and also by the observation from sediment cores that ocean productivity was highest at the onset of glacial conditions, when wind-blown dust was contributing iron to the ocean. But fertilizing the ocean with iron may not be a cure for the enhanced greenhouse effect as most of the carbon dioxide removed by the plankton seems to get recycled back into solution as they die or are eaten.

Ocean margins

Continents are often fringed by a shallow shelf, little more than 200 metres deep. Geologically, this is effectively part of the continent not the ocean and, at times of much lower sea level, parts of it must have been dry land. The continental shelves are often highly productive and support fisheries, or at least they did until over-fishing started to limit the catch. The organic productivity together with huge quantities of silt, mud, and sand washed by rivers or blown by wind from the neighbouring land has built up thick sediments. Where rivers supply these, the dense, sediment-laden water sometimes continues to flow almost river-like through gorges and over the edge of the continental shelf, sometimes continuing, as in the case of the Amazon, hundreds of kilometres offshore before dispersing in a delta-like pattern. In some places, the shelf margins have spectacular underwater scenery of cliffs and gorges that can only be seen by sonar but are as spectacular as any on land.

The ocean floor

Huge areas of the deep ocean floor are relatively flat and featureless, with little more than the occasional sea cucumber (actually a type of echinoderm, a relative of starfish) for many miles. But there are also mountains and canyons. We'll come to the mid-ocean ridges and trenches later, but there are also many isolated seamounts, sometimes known as guyots, rising from the ocean floor. Literally like underwater mountains, these are often isolated volcanoes, supplied in the past by mantle plumes even though they are not at the margin of a tectonic plate. Many are under more than 1,000 metres of water but carry evidence that they were once volcanic islands that rose above the sea, were eroded flat by waves, and subsided either individually or regionally back to the depths. Sometimes the subsidence was slow enough for coral reefs to build up around the island, leaving, after the volcanic land is gone, a circular atoll. Sometimes there are chains of the islands, formed as the ocean floor moves across a mantle plume. The most famous chain makes up the Hawaiian Islands and the Emperor seamounts to the northwest of Hawaii.

Landslides and tsunamis

The steep margins of the continental shelf and of seamounts means that the slopes can easily become unstable. There is evidence on the sea bed and on surrounding coasts of vast underwater landslides in which slope failure sends many cubic kilometres of sediments cascading down to the abyssal plain. Well-studied examples lie in the Atlantic, west of the islands of Madeira and the Canaries; off the northwest African coast; and off the coast of northern Norway. Sometimes the slide can be triggered by an earthquake, at other times it is simply that sediments have piled up too steeply and the slope fails. Either way, the underwater slides can generate devastating waves called tsunamis. There is evidence of three exceptionally large underwater slides during the past 30,000 years in the Norwegian Sea northwest of Norway. In one, about

7,000 years ago, 1,700 cubic kilometres of debris slid down the continental slope to the abyssal plain east of Iceland. The resulting tsunami flooded parts of the Norwegian and Scottish coastline to a depth 10 metres above the sea level at the time. An even more devastating slide occurred about 105,000 years ago south of the island of Lanai in Hawaii. That island experienced flooding 360 metres above the sea level of the time and the tsunami crossed the Pacific to deposit debris 20 metres above sea level in eastern Australia.

The sediment released by these huge slides and by the more gentle trickle of small slides down continental margins is buoyed up by water in a turbulent flow that can spread it considerable distances. It produces characteristic sediments called turbidites in which the grain size is graded in individual flows. The original slide might contain a mixture of grains but, as the flow fans out, coarse sand falls out more quickly than fine silt and mud, so individual flow bands will have a grading from coarse to fine within them. Such turbidites are often found in sequences of deep-water sedimentary rocks today.

Sea level

One of the clearest features on the surface of our planet is the boundary between land and sea: the coastline. It is one of the most dynamic environments on Earth, with features ranging from high rocky cliffs to low sand dunes and mud flats. And it is an environment into which, for some strange reason, large numbers of humans seem to flock in hot weather. But coastlines do not stay still. Some are eroding as the sea scours away millions of tons of material. In other places land is building up as the sea drives up sand banks or rivers extend their muddy deltas. Over geological timescales, the variations have been spectacular. In some cases huge parts of continents have been flooded in what are called marine transgressions. At other times the sea retreats – marine regressions. These apparent changes in sea level can be due to a number of reasons. One of the present concerns about global warming is that it might cause a rise in sea level. That can be due simply to the oceans warming so the water expands slightly; this alone could raise sea level by maybe half a metre in the next century. It could rise much further if there is significant melting of the Antarctic ice cap. (Melting of the Arctic ice and Antarctic sea ice would have no overall effect on sea level since the ice is already floating and thus already displacing its own weight in water.)

But all this is nothing compared to past sea level changes. Since the peak of the last Ice Age, sea level appears to have risen by as much as 160 metres. It fluctuated dramatically with the climate during ice ages over the last 3 million years. Going back further, sea level was at its highest between about 95 and 67 million years ago, during the upper Cretaceous, when shallow seas covered large areas of continents and thick deposits of chalk were formed together with many of the deposits that are yielding oil today. One theory for these exceptionally high sea levels is that large areas of ocean floor were being uplifted by hot material rising in the mantle as the Atlantic Ocean began to open. The geological record of sea level is characterized by periods of steadily rising seas followed by an apparently abrupt fall in sea level. Sometimes apparent sea level falls can be linked to the tectonic uplift of continents. In some instances it seems to happen globally and not always at the onset of an ice age. Perhaps it is sometimes due to sudden large-scale rifting in the ocean floor, literally pulling the floor out from under the sea.

Drilling the seas

From 1968 the ocean floor was sampled directly and scientifically by the US-led Deep Sea Drilling Program using a drilling vessel named the *Glomar Challenger*. This was superseded in 1985 by the international Ocean Drilling Program using the improved *JOIDES Resolution*. Around 200 separate two-month-long voyages or legs have taken place, with core samples drilled at intervals along each.

The deepest holes exceed 2 kilometres, and overall thousands of kilometres of core samples have been recovered. Many of them include varying depths of sediment down to volcanic basalt beneath. They all tell stories about their origins and the changing conditions of climate and ocean. Far from eroding land and river deltas, sedimentation rates are much lower. At high latitudes the sediments include clay and rock fragments rafted on icebergs which have melted and dropped their load. Elsewhere, wind-blown dust from deserts and volcanic ash makes up a greater proportion of deep-water sediment, sometimes accompanied by micro-meteorite dust, sharks' teeth, and even the ear bones of whales.

8. The ocean drilling ship *JOIDES Resolution*. The derrick towers 60 metres above the water-line.

Where ocean productivity is high on the surface, there are often also the sunken remains of various types of plankton. In comparatively shallow water, the calcareous skeletons of coccolithophores and foraminifera are common, forming a calcareous ooze that may consolidate to form chalk or limestone. But the solubility of calcium carbonate increases with depth and pressure. Somewhere between 3.5 and 4.5 kilometres down in the water, we reach the so-called carbon compensation depth (CCD), below which the tiny skeletons will tend to dissolve away. Here, their place can be taken by silicaceous ooze, made from the tiny silica skeletons of diatoms and radiolarians. Silica too would be soluble, but enough gets through to form significant deposits in the Southern Ocean and parts of the Indian and Pacific oceans. In a few places, usually where ocean circulation is restricted, such as in the Black Sea, the bottom water is anaerobic and black shales are deposited. They are sometimes rich in organic material that is not oxidized or consumed in the anaerobic conditions and can slowly turn into oil. Occasionally, the anaerobic deposits are more widespread, reflecting so-called anoxic events, where changes in ocean circulation prevent oxygen-rich waters sinking to the ocean floor.

Messages in the mud

Sediment cores can carry a long and continuous record of past climate. The types of sediments can reveal what was going on on the surrounding land – for instance, material rafted on icebergs or blown from deserts. But more precise records are kept by the ratios of stable isotopes of oxygen in calcareous ooze. Oxygen in water molecules exists in different stable isotopes, principally ^{16}O and ^{18}O. As sea water evaporates, molecules containing the lighter ^{16}O evaporate slightly more easily, leaving the sea water enriched in ^{18}O. That is soon diluted again by rainfall and rivers, except when large amounts of water get locked up in polar ice caps. Then the carbonate taken up by plankton and deposited in sediments will contain more ^{18}O than during the interglacial periods, so the oxygen isotopes in sediments reflect the global climate. By matching up the changes recorded in sediments for a total of more than 20 million years, the ocean drilling programme has shown how the climate fluctuates on timescales that seem to reflect the Milankovich cycle, the wobble of the Earth's axis, and the eccentricity of our orbit around the Sun.

In the 1970s the ocean drilling programme came to the Mediterranean. There, the drill cores reveal something sensational. I was shown one of them where it is now stored at the Lamont Doherty Geological Observatory of Columbia University in New York. It consists of layer after layer of white crystalline material, a mixture of salt (sodium chloride) and anhydrite (calcium sulphate). These evaporite layers can only have been formed by the Mediterranean drying up. Even today, evaporation rates are so high that, were the Straits of Gibraltar sealed off, the entire Mediterranean would evaporate in about 1,000 years. The implication of the hundreds of metres of evaporite in the drill cores are that this must have happened perhaps 40 times between 5 and 6.5 million years ago. When the scientists drilled close to the Straits of Gibraltar, they encountered a chaotic mixture of boulders and debris. This must have been the giant plunge pool of the world's greatest waterfall, when the Atlantic broke through past Gibraltar to refill the Mediterranean. We can only imagine the roar, the spray, the power of the water.

One of the most interesting of the recent legs of the ocean drilling programme involved drilling into deposits of gas hydrate. These are sediments containing high concentrations of methane ice, held in solid form by the low temperatures and high pressures of the deep ocean floor. There is added excitement when a gas hydrate core is returned to the surface as they can easily turn into gas again, sometimes explosively. That has made studying them somewhat difficult, but there are believed to be vast deposits of them. It is possible that they could become an economically important source of natural gas in the future. There are suggestions that they played a significant role in sudden climate change in the past. They can be quite unstable, and an earthquake can make large quantities float free of the ocean floor to rise up in great gas bubbles to the surface. A sudden fall in sea level can also destabilize gas hydrates so that they release their methane, which is a powerful greenhouse gas. It is probable that a sudden global warming 55 million years ago was caused by methane released from gas hydrates. It has even been suggested that some accounts of ships lost in the imaginary Bermuda Triangle in recent times originated from descriptions of large gas bubbles breaking the surface, capsizing boats or asphyxiating their crew.

Large quantities of organic material can become buried in ocean sediments and can, in the right circumstances, get turned into oil. That tends to happen in shallow marine basins that are undergoing crustal stretching. This thins the crust, deepening the basin so that it fills up with more sediment. But at the same time the organic material gets buried deeper, closer to the internal heat of the mantle so that it is cooked into crude oil and natural gas. This can then rise up through permeable strata and collect beneath impervious clays or salt layers. Rock salt is particularly mobile as it is not very dense and tends to rise up through strata in big domes. Often these can trap rich oil and gas deposits, as happens in the Gulf of Mexico.

Life underground

But not all the organic material in ocean sediments is dead. Living bacteria are often abundant in sediments more than 1,000 metres under the sea floor, in rocks over a hundred million years old. It seems likely that they were living in the sea floor mud and remained as it was buried deeper and deeper all that time ago. They don't exactly lead exciting lives but they're certainly not dead. It is estimated that they may divide only once every

1,000 years and live by anaerobically digesting organic material and releasing methane. Some bacteria can also survive at high temperatures, possibly up to the 100 to 150 degrees Celsius at which oil forms, and they may play a significant part in this process. It is possible that 90% of all terrestrial bacteria live underground and together comprise as much as 20% of the total biomass on Earth.

The longest mountain chain on Earth

If you were to drain the water from the world's oceans and reveal the spectacular landscape down there, the biggest feature would not be the great ocean island mountains taller than Everest or the great chasms that dwarf the Grand Canyon, it would be a mountain chain 70,000 kilometres in length: the mid-ocean ridge system. The ridges run around the planet like the seam in a tennis ball. Peppered along their length are volcanic fissures. Sometimes these erupt slowly underwater, producing pillow-shaped clumps of dense black basalt lava, like toothpaste from a tube. These are the zones of creation where new ocean crust is forming as the sea floor spreads.

The North Atlantic Ridge was discovered in the mid-19th century by a ship attempting to lay the first transatlantic cable. The ridges are broad, between 1,000 and 4,000 kilometres wide, and rise slowly towards a central line of peaks, typically 2,500 metres above the deep ocean floor but still a further 2,500 metres below the sea surface. The ridge is offset by numerous transform faults perpendicular to its length, displacing the ridge crest by many kilometres. The crest of the ridge often consists of a double line of peaks with a central rift between them. In the first half of the 20th century, proponents of the theory of continental drift such as Arthur Holmes suggested that the ridges might mark places where convection in the mantle brought new crust to the surface, but it was magnetic surveys which finally confirmed one of the most important discoveries in geology: sea floor spreading.

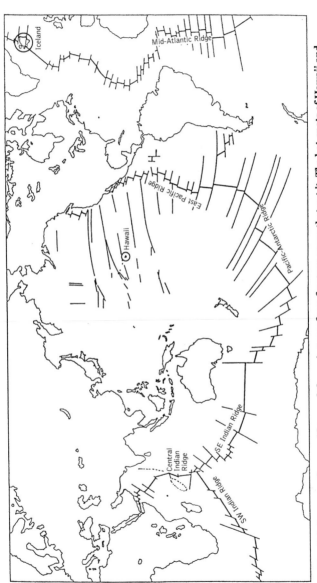

9. The global system of ocean ridges and the main transform fracture zones that cut it. The hot-spots of Hawaii and Iceland are circled.

Magnetic stripes

In the 1950s the US navy needed detailed maps of the ocean floor to aid their submarines. So research vessels began sailing to and fro making sonar measurements. Scientists were given the chance to contribute other experiments, and so it was that a sensitive magnetometer was towed across the oceans, mapping out the magnetic field. The map showed a series of highs and lows in the field strength that appeared like parallel stripes on either side of the mid-ocean ridges. It was Fred Vine and Drum Matthews at Cambridge who were able to confirm what was happening. As volcanic lava erupts and cools, it traps magnetic mineral grains aligned with the Earth's magnetic field. So, sail over recent submarine basalts and the magnetic field of the Earth will be slightly enhanced. But, as we heard in the last chapter, the Earth's

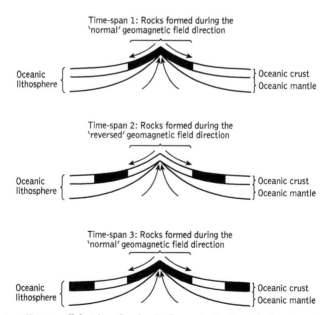

10. How parallel stripes develop in the magnetization of volcanic rocks on the ocean floor, as new ocean crust spreads out from an ocean ridge.

magnetic field sometimes reverses. Magnetism trapped in volcanic rocks that erupted when the field was reversed will carry an opposite component to the present field, lowering the reading slightly. Thus the magnetic stripes on either side of the mid-ocean ridge build up, with older and older sea floor as you move away from the central ridge in either direction. The sea floor is indeed spreading.

Boundary of creation

Overall, the spreading rate is slow but relentless, ranging from about 10 centimetres a year in the Pacific to 3 or 4 centimetres per year in the Atlantic, about the same rate at which your fingernails grow. But the eruption of lava to create new crust is not steady, which is why parts of the ridge get rifted and subside as they are stretched open and others build up peaks. Beneath the centre line of the ridge, hot mantle material is rising in a mush of crystalline rock which is partially melting. Along this line, the hot, soft asthenosphere rises to meet a thin ocean crust with no hard lithospheric mantle in between. Because this mantle material is hot it is less dense and so makes the ridge rise. About 4% of the mantle rock melts to form the basalt magma which percolates up through pores and fissures into a magma chamber a kilometre or so beneath

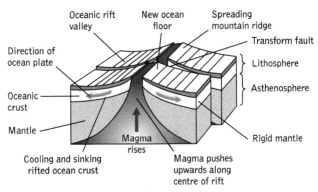

11. **The principal components of a mid-ocean ridge.**

the ridge. Seismic profiles reveal magma chambers several kilometres wide under parts of the Pacific ridge, though they are harder to see under the Atlantic ridge. Material in the magma chamber is slowly cooling, so some crystallizes out and accumulates at the bottom of the chamber to form a coarse-textured rock called gabbro. The remaining melt periodically erupts from fissures along the ridge. It is quite fluid and doesn't contain much gas or steam, so the eruptions are fairly gentle. But the lava is rapidly quenched by the sea water and tends to form into a series of pillow-like structures.

Black smokers

Even where there is not an active eruption, the rocks close to the ridge are still very hot. Sea water gets drawn into cracks and pores in the dry basalt, where it is heated and dissolves minerals such as sulphides. The hot water then rises out of vents, precipitating sulphide to form tall, hollow chimneys. Bacteria that can withstand the hot water contribute to the process by reducing soluble sulphates to sulphides. As the sulphides come out of solution in the cooling water, they form a cloud of black particles, so these vents are often known as black smokers. Water can gush out of them at high velocity and at temperatures in excess of 350 degrees Celsius, making them hazardous but fascinating to explore in deep-diving submersibles. The mineral chimneys can grow at a rate of several centimetres per day until they collapse in a pile of debris. In this way, considerable deposits of potentially valuable sulphide minerals can build up. Where the water is more acidic but slightly cooler, more zinc sulphide is dissolved, creating a white smoker. These are slower growing and generally cooler and thus better habitats for some of the amazing life forms that cluster around such hydrothermal vents. Life down here is based entirely on chemical energy, not sunlight. Primitive bacteria flourish in the hot and often acidic conditions. Blind shrimps, crabs, and giant clams feed off them, and giant tube worms containing symbiotic bacteria filter nutrients from the water. It has been suggested that life on Earth first began in such places, so they are causing much excitement among researchers.

Wealth from the sea

One of the surprise discoveries on the *Challenger* voyages in the 1870s was the return, in dredge samples from the deep ocean floor, of strange black nodules. These nodules are especially rich in manganese and iron oxides and hydroxides, together with potentially valuable metals such as copper, nickel, and cobalt. Known as manganese nodules, they are now known to pepper large areas of the deep ocean floor. Exactly how they form is uncertain, but it seems to be in a slow chemical process with the metals derived from sea water and possibly the underlying sediments. The nodules often grow in concentric onion-like layers around a small, solid nucleus, perhaps a chip of basalt, a bead of clay, or a shark's tooth. Estimates of their age suggest that they are very slow-growing, perhaps adding only a couple of millimetres in a million years. In the 1970s there were various proposals to mine them with scoops or suction, but so far this has not happened due to technological, political, ecological, and economic hurdles.

Pushes, pulls, and plumes

It does not seem as if sea floor spreading is the result of the ocean floor pushing apart from the mid-ocean ridge system. For most of their length, the ridges do not have substantial mantle plumes of hot material rising beneath them. It seems more as if they are being pulled apart, with new material rising to fill the gap. Beneath the ridge there is no thick, hard lithosphere, just a few kilometres of ocean crust. As mantle material rises under the ridge, the pressure drops and so does the melting point of some of the minerals. This leads to partial melting of as much as 20 or 25% of the material, producing basalt lava. The rate of magma formation is just right to produce ocean crust of a fairly uniform thickness of 7 kilometres.

A notable exception is Iceland, where a mantle plume and a mid-ocean ridge coincide. Here, far more basalt is erupted and the crust is around 25 kilometres thick, so that Iceland rises above the Atlantic. The history of that mantle plume can be traced in thickened basalt ocean crust right across the north Atlantic between Greenland and Scotland. Seismic surveys reveal that there are about 10 million cubic kilometres of additional basalt there, several times the volume of the Alps, or enough to cover the entire USA with a layer one kilometre thick. A lot of it didn't erupt on to the surface but was injected beneath the crust, under-plating it. The Hatton bank off the coast of Greenland is a bulge caused by such injections of basalt. The mantle plume that is now under Iceland may have been what caused the north Atlantic to begin to open about 57 million years ago. The volcanic activity appears to have started with a series of volcanoes, some of which are still preserved in the Inner Hebrides and Faeroe Islands northwest of Scotland.

Where oceans go to die

Ocean crust is constantly forming. As a result, it is difficult to find any truly ancient ocean floor. The oldest dates back to the Jurassic, about 200 million years ago, and is in the western Pacific. A segment about 145 million years old was recently discovered near New Zealand. But such ages are rare; most of the ocean floor is less than 100 million years old. So where have all the ancient oceans gone?

The answer lies in a process called subduction. As the Atlantic widens, the Americas on one side and Africa and Europe on the other are slowly moving apart. But the Earth is not getting bigger overall, so something must be taking up the slack. It appears to be the Pacific. The Pacific seems to be ringed by great trenches, up to 11,000 metres deep. Behind them is a ring of volcanoes on islands or on continents, the so-called Pacific ring of fire. Seismic profiling shows how the ocean plate – the thin ocean crust and as much as 100 kilometres of mantle lithosphere beneath – is plunging back

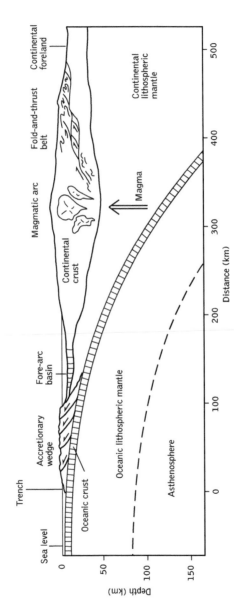

12. How ocean lithosphere subducts beneath a continent, accreting sediments along the margin and producing volcanic activity inland.

down into the Earth. In the 100 million years of its existence, the rocks of the lithosphere have steadily cooled and contracted, becoming more and more dense so that they can no longer float on the asthenosphere. It is this process of subduction that is one of the driving forces of plate tectonics: a pull rather than a push.

The cold, dense rock which sinks in a subduction zone has been under the sea, so it is wet. There is water in the pore spaces and also bound chemically in minerals. As the slab sinks and the pressure and temperature increase, the presence of water lubricates its flow but also lowers the melting point of some of its components, which rise through the surrounding crust to feed the ring of fiery volcanoes. As we saw in the last chapter, the rest of the slab of lithosphere continues down into the mantle, at least to the 670-kilometre boundary with the lower mantle, but eventually sinks perhaps as far as the base of the mantle. Seismic tomography can help trace its billion-year journey.

There are several different types of boundary between the slabs of continent and ocean lithosphere that make up the tectonic plates of the Earth. In the ocean there are the constructive boundaries of ocean ridges and the destructive ones where subduction occurs. This can take place where ocean lithosphere dives down beneath a continent, as in the case of the west coast of South America, forming the volcanic peaks of the Andes. Or ocean can dive beneath ocean, as with the deep trenches of the western Pacific, where the ring of fire comprises volcanic island arcs. There are boundaries where one plate grinds its way alongside another, such as along the coast of California. And there are also plate boundaries where continent runs into continent, but we will discuss them in the next chapter.

What's left on land

Not everything vanishes with a lost ocean. Where ocean lithosphere dives beneath the continents or where an entire ocean is squeezed out between two land masses, much of the sediments get scooped up and added to the continents. That is one of the reasons why so many marine fossils are to be found on land. Occasionally, entire masses of ocean crust can get lifted on to land, a process called obduction. Because they are from collision zones, such rocks are often very distorted, but, by piecing together the evidence from several such sequences, an overall picture emerges. They are known as ophiolite sequences, from the Greek for 'snake rock'. The descriptive name is also reflected in the name serpentinite, given because of the wiggly lines in the green minerals of the forms metamorphosed by hot water. At the top of an ophiolite sequence are the remains of ocean sediments followed by pillow lavas and sheets of basalt that may have been injected underground. Then comes gabbro, the slow-cooled crystalline rock of the same composition as basalt, and at its base the layered deposits of crystals from the bottom of the magma chamber. Beneath that there can be traces of the mantle rock from which the basalt was derived.

Lost oceans

Over hundreds of millions of years it is clear that many oceans have both opened and closed. For a long time, from 1,200 to 750 million years ago, the continents were clustered into one giant super-continent, surrounded by a single vast ocean spanning two-thirds of the globe. In the late Pre-Cambrian, the super-continent broke up into separate land masses. New oceans formed. One of them, the Iapetus Ocean, lasted between about 600 million and 420 million years ago. The join, or suture, where it closed again can today be crossed in a short drive across northwest Scotland. Half a billion years ago that journey would have involved a 5,000-kilometre sea crossing. By the Jurassic period of about 200 million years ago, a great wedge of ocean, the Tethys, had opened up between western Europe and southeast Asia where it opened into the Pacific. That closed as Africa hinged round into Europe to form the Alps and India came crashing into Tibet, lifting the Himalayas. Seismic studies trace remnants of the Tethys ocean floor descending into the mantle.

Over geological time, there have been numerous occasions when new oceans might have formed but did not. East Africa's Great Rift Valley and the Red Sea and Jordan Valley are obvious recent examples. The stretching of the North Sea basin, which produced North Sea oil deposits and Bavarian hot springs, is another. Another few hundred million years and our ocean charts will be completely out of date again.

Chapter 5
Drifting continents

As a child I used to enjoy helping my mother to make marmalade. I confess that I still like making it occasionally myself. But now, when I stare into the preserving pan of simmering fruit and sugar,

I can't help imagining that I am seeing our planet's evolution, greatly speeded up with one second perhaps representing ten or even a hundred million years. When the jam is gently simmering on a slow heat, convection cells establish, with columns of hot marmalade rising to the surface and spreading across it. With them comes some scum, a fine sugary foam, which is not dense enough to sink back down but collects in rafts on the calmer areas of the surface. This foam is a bit like the Earth's continents. It starts to form quite early on in the process and slowly builds up and thickens. Occasionally, the convection pattern beneath changes and the scum splits apart. Sometimes the rafts of scum run together and pile up even thicker. Of course, we shouldn't take this analogy too far. The timescales and the chemistry are altogether different; by and large geologists don't find sugar crystals in granite or orange peel xenoliths in basalt. But it's an image worth holding in mind as we consider the scum of the Earth: the continents.

Scum of the Earth

Continental crust is very different from the crust that floors the oceans. Ocean crust is predominantly magnesium silicate, whereas the continents contain higher proportions of aluminium silicates. They also contain less iron than the denser material of the mantle or of the ocean floor. As a result, they float, albeit on the semi-solid mantle rather than in liquid. And they can be thick. The ocean crust is a fairly uniform 7 kilometres thick, but the continents can range from 30 to 60 kilometres or more. And, like the ocean lithosphere, they are under-plated by a thick layer of cold, hard mantle. Just how deep the roots of continents do go is still a subject of controversy that, in the end, probably comes down to definitions. But continents are also a bit like icebergs: there's a lot more below ground than we can see above. And the higher they rise in mountain ranges, by and large the deeper they go beneath.

Drifting continents

With the benefit of hindsight, the knowledge of mantle convection, and the evidence of sea floor spreading, it is very easy to see that the continents have moved over geological time relative to one another. But it was not always so convincing. In spite of James Hutton's ideas about mountain-building and the rocks cycle, it was a long time before any mechanism could be suggested. Between 1910 and 1915, the American glaciologist Frank Taylor and the German meteorologist Alfred Wegener proposed the hypothesis of continental drift. Yet no one could imagine a way in which the continents could drift like ships at sea through the seemingly solid, rocky mantle. For nearly half a century, supporters of continental drift were in the minority. But the theory's few supporters were working hard. Alex du Toit in South Africa was building up evidence of similar rock structures between southern Africa and South America, while Arthur Holmes, a British geophysicist, proposed mantle convection as a mechanism for the drift. It was not until the 1960s, when the oceanographers got to work, that the debate was settled. Harry Hess proposed that convection beneath the ocean crust might cause the sea floor to spread out from mid-ocean ridges, and Fred Vine and Drum Matthews provided the magnetic evidence of sea floor spreading. It was papers by Tuzo Wilson in Canada, Jason Morgan at Princeton, and Dan McKenzie at Cambridge that brought the evidence together into the theory of plate tectonics.

Plate tectonics explains the surface of the Earth in terms of the motions of a small number of rigid plates which move relative to one another, interacting and deforming along their boundaries. It is not that the continents are drifting free but that they are carried on plates which extend far deeper to include the mantle lithosphere, typically 100 kilometres thick. The plates are not restricted to the continents, but include the slabs of ocean floor as well. There are seven principal plates: the African, Eurasian, North American, South American, Pacific, Indo-Australian, and Antarctic plates.

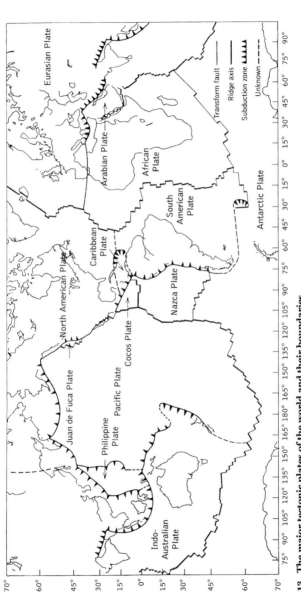

13. The major tectonic plates of the world and their boundaries.

There are also a number of smaller plates, including three quite substantial ones around the Pacific, plus some more complicated fragments where other plates join.

Another of my childhood memories is of tracing the continents from an atlas, cutting them out and trying to fit them together as a single land mass. This must have been about the time of Tuzo Wilson's paper in *Nature* in 1965. I can still remember the thrill of finding how well they fitted and of discovering some of the reasons why the fit was not perfect. It was not just down to my inaccurate tracings. As any nerdish schoolboy knows, you have to cut the continents at the edge of the continental shelf rather than at the coastline. And you can cut off the Amazon delta which would otherwise overlap with Africa, since that has grown since the split of the continents. More exciting was discovering that North and South America need to come apart to make a fit, and that Spain must part company from France. Swinging it back bangs Spain into France exactly where the Pyrenees are today. So could such continental collisions be the cause of mountain ranges?

It was at about that time in my teenage years that family holidays took me to the Pyrenees and to the Alps. In places I could see the layers of sedimentary rock not lying flat as they did in less disturbed lands but rucked up like a carpet into folds and undulations. This took my thoughts back to marmalade. As the conserve simmers, you keep a china plate in the fridge. Every few minutes you bring it out and drip a few drops of the hot marmalade on to it. When it has cooled, you push a finger into it. If it is still liquid, there's nothing for it but to lick your finger and let the marmalade continue to simmer. But after a while, as the brew approaches its setting point, a sample on the plate will crinkle up as you push your finger into it like a miniature continental collision. And it's not a bad model for the way continents behave on large scales. Compressed to fantastic pressures by overlying rocks, and possibly heated from beneath, rocks subjected to the lateral force of a colliding continent will tend to fold rather than to fracture. And the incredible masses of rock involved will be strongly affected by gravity so that the steepest folds will sag under their own weight into over-folds, rather as the skin on custard, or indeed marmalade, would do.

The Earth is not flat

Another reason why flat continents cut from an atlas do not fit together very well is that they are supposed to represent plates on the surface of a sphere. They get distorted in the map projection. But it is not easy to slide rigid plates about on the surface of a sphere either. You cannot simply move them in straight lines because there are no straight lines on a sphere. Each motion is in effect a rotation about an axis cutting through the sphere. But there are still difficulties. One is finding a frame of reference among all the jostling plates. Another is accommodating different rates of sea floor spreading. A simple model might invoke an axis similar to the Earth's rotation axis for the opening of the Atlantic and the relative motions of the Americas away from Africa and Europe. But that would require creating Atlantic Ocean crust like a segment of orange skin, wide at the equator and narrowing smoothly towards the poles. The rate of sea floor spreading does vary, but not in a convenient way like that. The result is transform faults; breaks in the crust thousands of kilometres long, offsetting segments of the mid-ocean ridge.

Frames of reference

With the evidence of sea floor spreading and the mechanism of mantle convection, plate tectonics rapidly became established at the centre of modern Earth sciences. But even today there are geologists who object to the term 'continental drift' because of its associations with the time when the mechanism had not been properly explained and few believed it. But, once people were prepared to accept it, evidence for past plate motions became obvious. There was the geological evidence of rocks of the same type split apart and now lying on opposite sides of an ocean. There was evidence from living and fossil remains of the times different populations became isolated from one another or when they were able to cross between continents. For example, it is just over 200 million years since Australia parted company from parts of Asia such as Malaysia and Indochina. Since then, mammals have been evolving independently on the two land masses, with the result that the marsupial line has come to dominate in Australia while placental mammals developed in Asia.

As with the evidence of magnetic reversals in ocean floor basalts that we discussed in the last chapter, so the magnetic evidence has provided the most comprehensive picture of past continental movements. The grains of magnetic mineral trapped like tiny compass needles in volcanic rocks when they solidified record the direction to the North Pole at the time. They show not only the small wiggles and big reversals in the magnetic field itself, but also trace out over tens or hundreds of millions of years, a larger, more sweeping series of curves: a so-called polar wandering curve. This is in effect the plot of how the continent itself has moved relative to the magnetic pole. When you compare the curves of different continents, sometimes you see that they move together but at other times they diverge, tracking how the continents themselves have split, drifted apart, and come together again in a sort of continental waltz. In fact, it's more like a clumsy barn dance, as the continents occasionally barge into one another.

Sensitive instruments even make it possible to track the relative continental movements today. Over short distances, such as locally across plate boundaries, surveying techniques and in particular laser ranging can be very accurate. But that can now be done over continental scales too, via space. Some of the strangest space satellites ever launched are for laser ranging. The satellite consists of a sphere of dense metal such as titanium with many glass reflectors, like cats' eyes, embedded in it. These reflect light back in the same direction from which it has come, so if you shine a powerful but narrow laser beam from the ground and time the reflected pulse back to where it started, you can work out the distance to an accuracy of centimetres. When you compare values from different continents you start to see how they are moving from one year to the next. Astronomers can do the same sort of thing with radio telescopes using distant cosmic sources of radio waves as their frame of reference. Now that the codes from the American military Global Positioning System (GPS) satellites are no longer scrambled, geologists can get similar precision using a small hand-held GPS receiver in the field. By the careful use of many readings, the accuracy comes down to millimetres. The answers confirm evidence of the rate of sea floor spreading: the plates are moving relative to one another at roughly the speed your fingernails grow, between about 3 and 10 centimetres per year.

Of course, all these plate motion measurements are relative to one another, and it is hard to establish a background frame of reference for them all. A clue comes from Hawaii. The Big Island of Hawaii is just the latest in a series of volcanic ocean islands that stretch away to the northwest and continue underwater as the Emperor seamount chain. Dates of the basalt reveal that the further northwest you go, the older the basalts are. It seems as if the chain marks the passage of the Pacific plate across an underlying plume of hot mantle material. Comparing the historical position of this with those of other mantle plumes shows little relative movement between them, so perhaps these mantle plumes are reference points in an underlying mantle which changes little. Estimates of absolute plate motions, relative to this framework, show that the western Pacific is moving the most. By contrast, the Eurasian plate is scarcely moving at all, so perhaps the historical choice of Greenwich as the reference point for longitude is geologically valid!

The continental waltz

Using a combination of geological and palaeomagnetic evidence, it is possible to trace back the motions of the tectonic plates through geological time. The continents we see today come from the

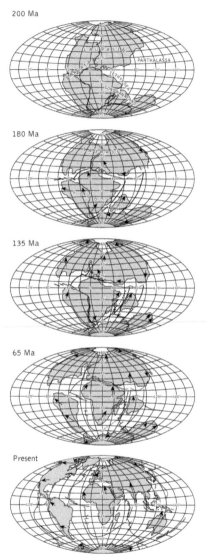

14. The changing map of the Earth's continents over the last 200 million years.

break-up of a super-continent that has been named Pangaea. That split apart in the Permian period about 200 million years ago, initially into a northern continent called Laurasia and a southern continent of Gondwanaland. The break-up of those continents continues today. But going further back in time, it seems that Pangaea itself was made up from an accumulation of earlier continents and that, still further back, there was an even earlier super-continent which has been called Pannotia, and one before that called Rodinia. These cycles of super-continent break-up, drifting, and re-accumulation have been called Wilson cycles, after Tuzo Wilson.

The further back in time you go, into the Pre-Cambrian, the less clear the picture becomes and the harder it is to pick out the land masses we know today. For example, in the Ordovician period, around 450 million years ago, Siberia was near the equator and most of the land masses were gathered in the southern hemisphere, with what is now the Sahara Desert near the South Pole. Late in the Pre-Cambrian, Greenland and Siberia were far south of the equator, Amazonia was almost at the South Pole, whilst Australia was well into the northern hemisphere.

One of the current record-holders for long-distance drifting must be the land known as the Alexander Terrane. It now forms a large portion of the Pan Handle of Alaska. About 500 million years ago, it was part of eastern Australia. The palaeomagnetic evidence in the rocks includes an inclination from the horizontal, dipping down into the Earth, which reveals the latitude at which the rock formed. The steeper the inclination, the higher the latitude. Other clues come from tiny grains of a mineral called zircon. These carry within them products of radioactive decay that date the periods of tectonic activity when they formed. For the Alexander Terrane, they reveal two major mountain-building episodes, 520 and 430 million years ago. Eastern Australia was the site of mountain-building at both these times, whilst North America was quiet. Conversely, western North America was active 350 million years ago, a time when the

Alexander Terrane seems to have been dormant. The Alexander Terrane began to split from Australia 375 million years ago and formed a submarine ocean plateau, at which time various marine animals were fossilized there. About 225 million years ago, the Terrane began moving northward at 10 centimetres a year. This continued for 135 million years, whereupon North American fossils start to appear as the Terrane reached its present latitude and collided with Alaska. It is even possible that it brushed past the coast of California on its way, scraping off material from the California Mother Lode gold belt. If that is correct, the Alaskan gold rush may have been to the same rocks as the Californian gold rush, but displaced 2,400 kilometres to the north.

Continental pile-up

We have heard about several different types of plate boundary. There are the spreading centres of mid-ocean ridges, the transform faults perpendicular to the ridges, and the subduction zones where ocean lithosphere takes a dive under a continent. All are comparatively narrow, well-defined zones that are relatively easy to understand and explain with simple diagrams. But there is one type of plate boundary that is more complex and where the idea of the tectonics of rigid plates breaks down: intercontinental collisions. Where ocean crust is involved it is comparatively easy. As long as it is cold, it will be dense enough to sink down into the mantle at a comparatively steep angle of about 45 degrees. Continental crust won't sink, just as a cork floating in the sea stays afloat in spite of all waves breaking over it. Continental crust also deforms more easily than ocean lithosphere. So, when continents collide, it is more like a serious traffic accident.

A good example is how India joined Asia. For hundreds of millions of years, India had been one of the partners in a complex country dance across the southern hemisphere featuring Africa, Australia, and Antarctica. Then, about 180 million years ago, it broke away and began to drift northwards. There are some spectacular mountain ranges down the western side of India, the Western Ghats and Deccan Traps. A strange feature of them is that, although sea is quite close to the west, the major rivers that drain these ranges flow to the east. Another puzzle came when Professor Vincent Courtilliot of the University of Paris started to look at palaeomagnetism in the basalt rocks of which the hills are made. He had been working in the Himalayas and wanted some comparisons from further south. He expected to find millions of years' worth of palaeomagnetic data in the thick basalt layers, spanning many palaeomagnetic reversals. Instead, he discovered that they were all magnetized in the same direction, suggesting that they must have erupted within one short period of at most a million years. Keith Cox at Oxford University and Dan McKenzie at Cambridge worked out what must have happened. India was once a larger continent, and as it drifted north, it passed over a mantle plume at just the time it was producing a huge pulse of magma. This caused the continent to dome up. But the Deccan Traps are just on the east side of this dome. Rivers could not drain to the west since that lay uphill. Unimaginable volcanic eruptions produced several million cubic kilometres of basalts in the space of thousands of years. Eventually, the activity split the continent in two. The Indian subcontinent we know today is just the northeastern portion of that. The rest lies under the sea in the huge basalt bank between the Seychelles and the Comoros. It turns out that this volcanic outpouring occurred around 65 million years ago, around the time of the Cretaceous/Tertiary boundary, and the extinction of many animal groups including the dinosaurs. Maybe it was not an asteroid after all that killed them but the pollution and climate change that resulted from these incredible volcanic eruptions.

Meanwhile, what remained of the subcontinent continued northwards, closing the great oceanic gulf of the Tethys and eventually ramming into Asia. Whilst the ocean lithosphere of the Tethys was dense and could subduct back into the mantle beneath Asia, the continental crust could not. The two continents first came into contact about 55 million years ago, but a continent has such momentum that nothing is going to stop it dead in its tracks. The closing speed was about 10 centimetres per year. It slowed to around 5 centimetres per year but has continued colliding ever since, like a vehicle crash test played in very slow motion. During this time, the Indian subcontinent has moved a further 2,000 kilometres north. The first thing to happen was a pile-up of sediments and a thickening of the crust in a series of under-thrusts as slabs of Indian continent wedged beneath Asia, like the debris in

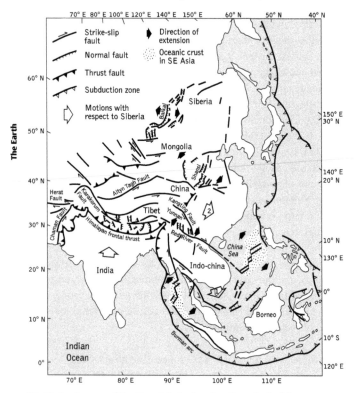

15. **Tectonic map of Southeast Asia showing the principal fractures resulting from the collision of the Indian subcontinent and the motions of China and Indochina as they are squeezed to one side.**

front of a bulldozer. This thick continental material gave rise to the high Himalayas.

As you head north across the flat Ganges plain, the first of these giant thrust faults forms a sharp feature. Here and there in the line of hills are sediments that were once deposited on the river bed but are now lifted tens of metres above it, yet they are only a few thousand years old, suggesting that there has been sudden and dramatic uplift in earthquakes. These hills are the foothills of the Himalayas which rise in a series of ranges stretched out to the east and west. Each line of mountain peaks corresponds approximately to another huge wedge of continental rock. The rocks exposed in the Himalayas today are mostly ancient granites and metamorphic rocks that have been uplifted from deep within the continents. The folded sediments scooped up from the floor of the Tethys Ocean lie north of the mountains on the edge of the Tibetan plateau. Behind them is a line of lakes corresponding to the original join between the continents, known as the suture.

An old, cold continent such as India is hard and relatively rigid. The part of Asia it collided with is relatively young and soft. Just as the hot mantle can undergo solid flow, so can crustal rock. We think of rock as we find it at the surface, hard and brittle, but crustal minerals such as quartz can flow like toffee at temperatures of only a few hundred degrees Celsius, just as olivine can deeper in the mantle. Some of the best models for the collision of India with Asia come if India is taken to be a relatively solid continent being driven into something with many of the properties of a liquid. And it is a liquid rather like non-drip paint – the harder you push on it, the easier it becomes to deform it. Models of this sort can account for the patterns of mountain ranges in central Asia but not for the high plateau of Tibet.

The rise of Tibet

The pile-up of relatively low-density crustal rocks could not simply be accommodated by downward thrusting, and the entire region began to float upwards. The dense lithospheric root beneath Tibet detached and sank back into the underlying asthenosphere. The remaining thickened continental rocks floated upwards, lifting the Tibetan plateau by as much as 8 kilometres. At the same time, parts of Asia tried to slide out of the way, with Indochina heading eastwards. This sideways motion stretched the continent further north, causing, among other features, lake-filled rifts in Tibet and the deep rift of lake Baikal in Russia. With some of the underlying cold, dense lithosphere removed, the hot asthenosphere was sufficiently close to the Tibetan crust to cause localized melting and account for the recent volcanic rocks found in parts of that country. There is also seismic evidence for a vast pool of partially molten granite about 20 kilometres under the southwestern part of the Tibetan plateau. That would also help to explain how Asia absorbed the impact of India and why the Tibetan plateau has remained relatively flat though surrounded by high mountain ranges. It seems that, overall, the Tibetan plateau is unlikely to get any higher than its present average of 5,000 metres. Any additional uplift would be balanced by a flow of material away to the sides. The more mountainous areas too have an average elevation of no more than about 5,000 metres above sea level. Here, heights are kept in check by erosion. Although a great deal of material has already been eroded from the Himalayas, regions such as Nanga Parbat in the north of Pakistan are still rising today by several millimetres per year, making slopes unstable and prone to landslides.

Monsoon

The Himalayas reach almost as high as passenger aircraft normally fly, and the mountains pose a significant barrier to atmospheric circulation. The result is that central Asia to the north remains cold in winter and dry for most of the year. In the summer, warm air rising above the Tibetan plateau holds back moist air from the southwest, so that the clouds build up and release their moisture in the torrential rain of the Indian monsoon. The monsoon stirs up the Arabian Sea, bringing nutrients to the surface and resulting in an annual plankton bloom. This in turn leaves its trace in the sediments beneath. Sediment cores show that this sequence began around 8 million years ago, perhaps corresponding to the end of the major uplift of the Tibetan plateau and the origin of the monsoon weather pattern. Wind-blown dust in China shows that the region north of the Himalayas was drying out at about this time too. There is also a change in the sediments off the west coast of Africa, with an increase in wind-blown dust in the layers. It seems that this corresponds to the start of the drying out of Africa and the beginnings of the Sahara Desert as the moist clouds were drawn away towards India. There is a theory that the huge amount of chemical weathering that must have taken place in the eroding Himalayas drew down so much carbon dioxide from the atmosphere that, in turn, it may have set the stage for the ice ages of the last 2.5 million years. So perhaps the climatic changes in Africa that provided the evolutionary pressures that led to the development of modern humans there also have their origins in the rise of Tibet and the Himalayas.

Swiss roll

Further west than the continental pile-up of the Himalayas, the Tethys Ocean narrowed to an inlet, but the results of the collision, in this case of Italy and the African plate with Europe, are similar, if on a slightly smaller scale. The Alps are one of the most studied and best understood mountain ranges. To the north lies a sedimentary basin which slowly filled with sediments known as molasse. South of the Alps, in Italy, lies the plain of the River Po, equivalent to the Ganges plain in India. Between it and the mountains is a series of wedges of sediment, scooped up from the Tethys Ocean, sediment known as flysch. Then come the high Alps of Switzerland, made of the crystalline base of the continent, together with intrusions of granite from partial melting beneath. Beyond them come a series of very strongly folded rocks scooped up into giant over-folds called nappes, folding to the north and sagging under their own weight, as if they had been scooped up like whipped cream. These nappe folds are often so extensive that older rocks are folded up above younger rocks in a very confusing sequence. As with the Himalayas, there is a series of thrust faults, in places doubling the thickness of the continental crust.

Cratons

No continent is an island entire of itself. Continents can split apart or join up and merge. Modern mountain ranges such as the Alps and the Himalayas are just the latest examples of this. Others are so ancient that they have been worn down almost flat again. The Caledonian range of northwest Scotland and the Appalachians of North America are examples dating from when a forerunner of the Atlantic closed about 420 million years ago. The modern continents are patchwork quilts of such features. But the older and thicker a continent becomes, the more rigid it grows and the longer it lasts. The most stable cores of continents that are least affected by tectonic movements are known as cratons, and they make up the cores of present-day North and South America, Australia, Russia, Scandinavia, and Africa. Over time, they can often undergo slow subsidence. Lake Eyre in Australia and the Great Lakes of North America occupy such basins. The southern African craton, by contrast, has been uplifted by the buoyant rocks of a mantle plume beneath.

Profile of a continent

The same sort of principle that revealed the structure of the Earth as a whole, seismic tomography, can study the deep interior of continents in great detail. To get the high resolution required, the technique relies not on random natural earthquakes on the other side of the world, detected by widely spaced seismometers, but creates artificial seismic waves and picks up their reflections using nearby, closely spaced arrays of detectors. It is very expensive and at first was the monopoly of the oil exploration companies, which jealously guarded the results. But now there are many national projects which are sharing their data. The most advanced of these are in North America, where the Consortium for Deep Continental Reflection Profiling in the USA and Lithoprobe in Canada have built up a detailed series of profiles. To create the shock waves they employ a small fleet of purpose-built trucks which use hydraulic rams to shake the ground with heavy metal plates. Deep vibrations are monitored by a network of sensors over many miles which record the reflections from numerous layers beneath the ground. Computer analysis reveals each discontinuity or sudden change in

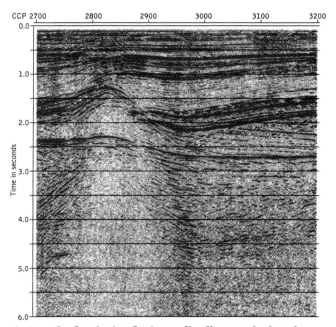

16. Example of a seismic reflection profile of layers and a domed structure within the Earth's crust.

density. These profiles go far deeper than the sedimentary basins of most interest to the oil prospectors. They reveal the ancient sutures between continents that merged with one another long ago. They have revealed reflections from a layer descending into the mantle beneath the Lake Superior region of Canada that could be the oldest subduction zone yet found, with the floor of a lost ocean about 2.7 billion years old. The profiles reveal how basalt magma rising from the mantle and unable to break through the thick continent under-plates it with sheets of basalts known as dykes. They also reveal how, when continental rocks get buried deep enough, they begin to melt so that they rise up through the continent to recrystallize as granite.

The rise of granite

As continental rocks pile up, so the base of the continent gets buried deeper and deeper. As it sinks, it heats up and the rocks at base begin to melt. Many of them are ancient sediments deposited in seas billions of years ago. They contain water chemically bonded into the rocks. The water helps them to melt and lubricates them so that they rise easily towards the surface. Unlike volcanic rocks, they're too thick and sticky to erupt from volcanoes. Instead, huge bubbles of molten rock, perhaps many thousands of metres across, push up into the higher layers of the continents, perhaps at quite high speeds. They bake the surrounding rock and cool slowly, forming a coarsely crystalline rock of quartz, feldspar, and mica: granite. Eventually, the surrounding rock wears away to reveal the great granite domes of, for example, Dartmoor.

Granite may be the inevitable result of a tectonically active planet of silicate rocks and plenty of water. But there can never be a Waterworld, a planet with no continents but global oceans. Once there is water, it finds its way into the chemistry of the rocks, lubricating them as they melt so that they can rise as great masses of granite to form the peaks of continents above the oceans. Without water, you get the situation on Venus: tectonics without the plates.

Without the inner fire of molten magma, you get the situation on Mars, an old, cold surface where life, if it exists, is deep in hiding. On Earth, you get oceans and continents in dynamic and sometimes lethal interaction.

Riches in the earth

One of the first incentives to geological exploration was the search for mineral wealth. Rare and valuable substances can be formed or concentrated by a number of geological processes. Organic remains in sedimentary basins can be gently cooked by the Earth's heat to produce coal, oil, and natural gas. We've already seen how sulphides of valuable metals can be concentrated around deep-sea hydrothermal vents and how manganese nodules can form on the deep ocean floor. Minerals can be concentrated in continental rocks in a number of ways. In molten rock, crystals will begin to form and the densest ones will sink to the bottom of the melt chamber. As well as concentrating minerals within it, a mass of molten rock rising through other rocks will drive super-heated water and steam ahead of it. Under pressure, that can dissolve many minerals, especially those rich in metals, forcing them through cracks and fissures where they are deposited as veins of mineral. Other minerals can become concentrated near the surface when water evaporates or when other components in a rock are eroded away. If we have the technology to recover them, the Earth's riches are there for the taking.

The search for lost continents

If continental scum has been accumulating on the surface of the planet for most of its history, when did it begin? Where is the first continent? It is not easy to say. The most ancient continental rocks have been so reworked, folded, fractured, buried, partially melted, folded and fractured again, and shot through by younger intrusions, that it is hard to make sense of them. It's a bit like trying to identify the remains of an individual car within the compacted scrap from a junk yard. But the search for the oldest rocks on Earth may be nearing its end. Some of the first contenders were from the Barberton greenstone belt in South Africa. These are more than 3.5 billion years old, but they are the remains of pillow lavas and ocean islands, not continents. Similar rocks have now been discovered in the Pilbara region of western Australia, and there are rocks in southwest Greenland that yield dates of 3.75 billion years, but these again are ocean volcanic rocks. The best candidate for the first continent lies in the heart of northern Canada. In the uninhabited, barren lands about 250 kilometres north of Yellowknife, close to the Acasta River, there stands a lonely shed filled with geological hammers and camping equipment. Above the door is a rough sign, 'Acasta City Hall, founded 4 billion years ago'. Some of the rocks from around there have yielded dates a fraction over 4 billion years old.

They have given up their secrets thanks to grains of the mineral zircon, which traps within its lattice uranium atoms, which decay into lead. The grains can be disturbed by re-melting, later growth, and cosmic ray damage, but an instrument developed in Australia known as a SHRIMP (a Sensitive High Resolution Ion Micro Probe) uses a narrow beam of oxygen ions to blast atoms off tiny portions of the zircon so that different zones of the grain can be analysed individually. The centres of some of the grains have given ages of 4.055 billion years, making them the oldest rocks on Earth, and evidence of continents less than 500 million years from the Earth's formation.

Eternity in a grain of sand

But there is tantalizing evidence for something even older. About 800 kilometres north of Perth in western Australia, in the Jack Hills, there are rocks of conglomerate, a mixture of rounded grains and pebbles bound into rock about 3 billion years ago. Among the grains within that rock are zircons that must have eroded out from even earlier rock. One of these has given an age of 4.4 billion years, and an analysis of oxygen atoms in the crystals suggests that the Earth's surface at that time must have been cool enough for liquid water to condense. This research suggests that there were continents far earlier than anyone had expected, within a hundred million years of the Earth's accretion, and seems to run contrary to the concept of a partially molten, inhospitable world at that time.

Super-continents of the future

We have spent most of this chapter looking back in time at the continental waltz of the past. But the continents are still on the move, so what will the world map look like in another 50, 100, or more million years? At first, it is reasonable to assume that things will continue in their present directions. The Atlantic will continue to widen, the Pacific will contract. The process which closed the Tethys Ocean will continue, with more earthquakes and mountain uplift in the hazardous country between the Alps and the Himalayas. Australia will continue north, catching on Borneo and twisting round to collide with China. Further into the future, some motions may reverse. We know that a predecessor of the Atlantic opened and closed in the past, and it is probably inevitable that the Atlantic Ocean crust will eventually cool, contract, and start to sink again, perhaps subducting under the east coast of the Americas. Then the continents will bunch up again. Christopher Scotese of the University of Texas in Arlington predicts that, 250 million years into the future, there will be a new super-continent, Pangaea Ultima, possibly with an inland sea, all that will remain of the once-mighty Atlantic Ocean.

Chapter 6
Volcanoes

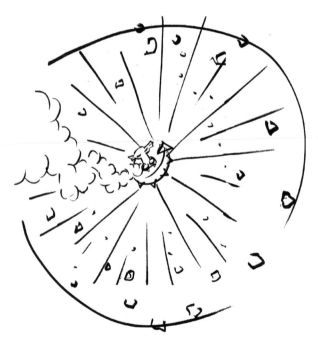

The relentless motions of tectonic plates, the uplift and the erosion of mountain ranges, and the evolution of living organisms are processes which can only be fully appreciated across the deep time of geology. But some of the processes at work in our planet can manifest all too suddenly, changing the landscape and destroying lives on a very human timescale: volcanoes. Superimpose a map of active volcanoes on a world map showing the boundaries of the tectonic plates and their association is obvious. The ring of fire around the Pacific, for example, is clearly associated with the plate boundaries. But where is the molten rock that feeds them coming from? Why are volcanoes different from each other, with some producing gentle eruptions and regular trickles of molten lava, whilst others erupt in devastating explosions? And why are some volcanoes, such as those of Hawaii, in the middle of the Pacific, far from any obvious plate boundary?

History is littered with eyewitness accounts of volcanic eruptions and with explanations, some of them mythological, some fanciful, and some surprisingly accurate. Among the better accounts is that of Pliny the Younger, describing the eruption of Vesuvius in 79 AD which killed his uncle, Pliny the Elder, and destroyed Pompeii and Herculaneum. But for a long time no one understood the cause of volcanic eruptions. Often they were thought to be the work of a fire god or goddess, such as the Hawaiian goddess Pelée. In medieval Europe, volcanoes were thought of as the chimneys of hell. Later, it was suggested that the Earth was a cooling star with remnants of the stellar fire within, linked through a system of fissures. In the 19th century, what we now know to be volcanic rock was widely thought to be deposited from oceans, the Neptunist theory, as opposed to the Plutonist idea that it had once been molten. After the Plutonist view gained ground, many thought that the interior of the Earth must be molten, an idea that did not get ruled out until the dawn of seismology. One of the mysteries was that volcanic rocks can have different compositions; sometimes even when erupted from the same volcano. Charles Darwin was one of the first to suggest that the composition of the melt could change as a result of dense minerals crystallizing out and sinking in the magma, something that he backed up by observations of volcanic rocks in the

Galapagos islands. As with his ideas on continental drift, it was Arthur Holmes in the mid-20th century who was the first to come close to the truth with his ideas of convection within the solid Earth's mantle.

How rocks melt

The key to understanding volcanoes comes from understanding how rocks melt. For a start, they don't have to melt completely, so the bulk of the mantle remains solid even though it gives rise to a fluid, molten magma. That means that the melt does not have the same composition as the bulk of the mantle. As long as the so-called dihedral angles, the angles at which the mineral grains in mantle rock meet, are large enough, the rock behaves like a porous sponge and the melt can be squeezed out. Calculations show how it will tend to flow together and rise quite rapidly in a sort of wave, producing lava at the surface in the sort of quantities seen in typical eruptions.

Melting does not necessarily involve increasing the temperature. It can result from decreasing the pressure. So a plume of hot, solid mantle material will begin to melt as it rises and the pressure upon it reduces. In the case of a mantle plume, that can happen at considerable depths. Helium isotope ratios in the basalt erupted on Hawaii suggest that it originates at around 150 kilometres depth. The mantle there is composed mainly of peridotite, rich in the mineral olivine. Compared to that, the magma that erupts contains less magnesium and more aluminium. It is estimated that as little as 4% of the rock melts to produce Hawaiian basalts.

Beneath the mid-ocean ridge system, the melting takes place at much shallower depths. Here there is little or no mantle lithosphere and the hot asthenosphere comes close to the surface. The lower pressures here can result in a larger proportion of the rock melting, perhaps 20 or 25%, supplying magma at about the right rate to sustain sea floor spreading and produce an ocean crust 7 kilometres thick. Most of the ocean ridge eruptions pass unnoticed as they take place more than 2,000 metres underwater as rapidly quenched pillow lavas. But seismic studies have revealed magma chambers a few kilometres beneath the sea floor in parts of the ridges, particularly in the Pacific and Indian oceans, though there is also some evidence of magma chambers beneath the mid-Atlantic ridge. Where a mantle plume coincides with an ocean ridge system, as in the case of Iceland, more magma is generated and the ocean crust is thicker, in this case rising above the sea to form Iceland.

Hawaii

The Big Island of Hawaii has welcoming people and friendly volcanoes. The town of Hilo is probably more at risk from tsunamis triggered by distant earthquakes than from the great 4,000-metre volcano of Mauna Loa that looms behind it. To the north and west lie the other Hawaiian islands and the Emperor seamount chain, tracing the long journey of the Pacific plate across the hot spot of an underlying mantle plume. To the south of the Big Island of Hawaii is Loihi, the newest of the Hawaiian volcanoes. As yet it has not broken the surface of the Pacific, but it has already built a high mountain of basalt on the ocean floor and will almost certainly become an island above water before long. Because Hawaiian lava is very fluid, it can spread over a wide area and does not tend to form very steep slopes. Such volcanoes are sometimes known as shield volcanoes, and they can flood basalt over a wide area. Often, a particular flow will develop a tunnel around it as the outer crust solidifies but the lava continues to flow inside. When the supply of lava ceases, the tunnel can drain and be left hollow.

The last big volcano to finish erupting on Hawaii, Mauna Kea, is home to an international astronomical observatory beneath some of the clearest skies on Earth. It was whilst visiting there one night that I saw, through binoculars, a distant fiery eruption on the flanks

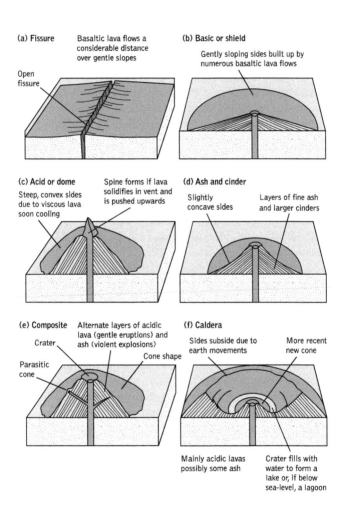

(a) Fissure Basaltic lava flows a considerable distance over gentle slopes

Open fissure

(b) Basic or shield

Gently sloping sides built up by numerous basaltic lava flows

(c) Acid or dome Spine forms if lava solidifies in vent and is pushed upwards

Steep, convex sides due to viscous lava soon cooling

(d) Ash and cinder

Slightly concave sides

Layers of fine ash and larger cinders

(e) Composite Alternate layers of acidic lava (gentle eruptions) and ash (violent explosions)

Crater

Parasitic cone

Cone shape

(f) Caldera

Sides subside due to earth movements

More recent new cone

Mainly acidic lavas possibly some ash

Crater fills with water to form a lake or, if below sea-level, a lagoon

17. The principal types of volcano, classified by shape (not to scale).

of Mauna Loa, in the Puu Oo crater. The next day I was able to take a helicopter flight low over the freshly erupted lava flows. Through the open door I could feel the radiant heat from the glowing, and in places still moving, lava and smell traces of sulphur in the air. But it all seemed quite safe, even hovering in the crater itself, though avoiding the plume of smoke and steam. The nearby Kilauea caldera, where many of the current eruptions take place, boasts an observatory and viewing platforms. Every few weeks, visitors can witness a new eruption, often beginning with a curtain of fire with many fountains of glowing magma along a rift. Nothing is burning in the fire, but the release of volcanic gases through the hot, runny lava causes incandescent streams to fountain tens or even hundreds of metres into the sky. The eruption may last only a few hours. In spite of the high fountains, eruptions are not particularly explosive due to the very runny nature of the lava. This allows vulcanologists from the nearby Volcanoes Observatory to approach the molten lava, and even the vents, wearing heat-protecting suits. They will probably have already detected the magma rising in the vent through a sensitive network of seismometers and by measuring the change in the pull of gravity brought about by the upwelling magma. Sometimes they are able to collect uncontaminated samples of volcanic gas directly from the vents and take the temperature of the molten lava. It erupts at about 1,150 degrees Celsius.

Plinean eruptions

The nature of a volcanic eruption depends on the viscosity, or stickiness, of the magma and also on the amount of dissolved gas and water it contains. Early in an eruption any groundwater will be flashed explosively to steam. As the gas comes out of solution when the pressure is released, it will expand rapidly, sometimes explosively. Moderate amounts of gas in runny basalt produces the fire fountains of Hawaii. Greater quantities of gas will carry with it finely divided solid material such as ash and cinders. Eruptions tend to be more violent early on when the magma still contains a lot of gas. If it has had time to settle in a comparatively shallow magma chamber, it will be better behaved. Sometimes the gas and ash rise so high into the air that they spread widely before the ash falls. This was the sort of eruption witnessed on Vesuvius in 79 AD and has been called a Plinian eruption after the account by Pliny of his uncle's death.

Many different volcanic rocks can be produced in such eruptions. Where the ash and cinders are hard before they land, layers of loose tuff will build up. If the fragments are still molten, it would be a welded tuff. Near the vents, larger lumps of magma will be thrown out. If they are still molten when they hit the ground, they will form splatter bombs that look a bit like cow pats. If a solid crust forms in the air around the still-expanding lava bomb, it will form a bread-crust bomb, looking rather like a loaf of risen bread. Lava that is rapidly quenched can form volcanic glass called obsidian. If the lava solidifies with gas bubbles still in it, these are known as vesicles. Sometimes a foam of gas bubbles in lava can form, creating pumice of such low density that it floats on water. The surface of a lava flow can be very rough and cinder-like, in Hawaii called 'aa' lava. (This is a Hawaiian word, not just what you say when you try to walk over it!) Where a thin skin forms on a fluid lava flow, it can crinkle into flow lines, creating ropy or 'pahoehoe' lava. Occasionally, fine strands of lava become drawn out, an effect sometimes known as Pelée's hair.

Ring of fire

As long as you keep clear of the fire fountains and the fast-flowing lava, Hawaiian eruptions are reasonably safe. But that is not true of most volcanoes. Much of the Pacific is circled by a ring of fire – volcanoes of a much more temperamental nature. Where an ocean plate dives down in a subduction zone beneath a continent or an island arc, so-called stratovolcanoes develop. These can be picture-postcard volcanoes. Mount Fuji in Japan is one, with steep, conical slopes and a snow-capped peak topped by a smoking crater. But the beauty of such mountains can mask their sinister behaviour. They are notorious for earthquakes and sudden, violent eruptions, as Mount Unzen in Japan and Mount Pinatubo in the Philippines both showed in 1991. They are called stratovolcanoes because of their stratified structure of alternate layers of lava and ash or cinders, which can spew over an area far wider than the peak itself.

The reason for their violent behaviour compared to Hawaiian volcanoes is that they are not fed by clean, fresh magma from the mantle. The material sinking down beneath them is old ocean crust. It is saturated with water, both as liquid in pores and fissures but also bound into hydrous minerals. As the slab descends, it heats, due to its depth and also possibly due to friction. The presence of water lowers the melting point, so partial melting occurs. The pressure is so great that the water easily dissolves in the melt, lubricating it, and this magma squeezes up through the continental crust above. As it nears the surface, the pressure drops and the water starts to escape as steam. It can do so very rapidly and violently, in much the same way as gas escapes when a well-shaken bottle of champagne is uncorked.

On its way up, the magma may accumulate in chambers until there is enough pressure to erupt. During this time, dense minerals can solidify and fall to the floor of the chamber. These minerals, particularly iron compounds, are what make basalts dark and dense. The melt that remains for the eruption is lighter-coloured and richer in silica – up to 70 or 80% silica in some cases, compared to 50% or less in basalt. It forms rocks known as rhyolites and andesites, typical of places like Japan and the Andes. Their eruption is violent, not only due to the high water content but also because such silica-rich lavas are much more viscous and sticky. They do not flow easily, and bubbles can't easily escape. Such lava cannot fountain like a Hawaiian eruption but is flung out of the way explosively.

Mount St Helens

One of the most famous eruptions of recent times was of a volcano of this type. Mount St Helens is in Washington State in the northwestern USA, where the Pacific plate is subducting. At the beginning of 1980 it was a beautiful mountain surrounded by pine forests and lakes, a popular holiday retreat. It had shown little activity since 1857. Then, on 20 March 1980, in a series of small earth tremors building up to a magnitude 4.2 quake, it reawakened. Earth tremors continued to increase and trigger minor avalanches until 27 March, when a big explosion occurred in the summit crater and Mount St Helens began to spew ash and steam. The prevailing wind blew the darker ash to one side, leaving white snow on the other. The mountain took on a two-tone appearance.

So far, no lava had erupted. All that escaped from the crater was steam, blowing ash with it. But it signalled the rise of hot magma beneath the volcano. The earthquake activity continued, but the seismographs also began to record a continuous rhythmic ground-shaking different from the sharp shocks of the earthquakes. This so-called harmonic tremor is believed to have been generated by magma rising beneath the volcano. By mid-May, 10,000 earthquakes had been recorded and a prominent bulge had developed in the north flank of Mount St Helens. By firing laser beams at reflectors placed around it, geophysicists were able to measure its growth. It was pushing northwards at an incredible 1.5 metres per day. By 12 May, certain parts of the bulge were more than 138 metres higher than before the magma intrusion began. The volcano was literally being wedged apart into a highly unstable and dangerous condition.

Early on the morning of Sunday 18 May, Keith and Dorothy Stoffel were in a small plane above the mountain when they noticed rock and snow sliding inwards into the crater. Within seconds, the whole north side of the summit crater began to move. The bulge had collapsed in a great avalanche. It was like taking out the cork from a

18. The eruption of Mount St Helens in Washington State in 1980 was one of the most spectacular and best documented in recent times. The column of ash and smoke rose nearly 20 kilometres into the atmosphere.

champagne bottle. The magma inside it was exposed. The explosion was almost instantaneous. The Stoffels put their plane into a steep dive to gain speed and escape. David Johnston of the US Geological survey was not so lucky. An hour and a half before he had radioed in the latest laser beam measurements from his observation post 10 kilometres north of the volcano. As the north flank fell away, the blast headed straight for him. He was one of 57 people to die in the eruption.

Though it began several seconds later, the blast quickly overtook the avalanche. It fanned out at more than 1,000 kilometres per hour. Over a radius of about 12 kilometres, trees were not only flattened but swept away. Nothing living or man-made was left. As far as 30 kilometres away, trees were toppled, though isolated pockets survived in hollows. Even further afield, the leaves were seared by heat and branches were snapped.

Soon after the lateral blast, a vertical column of ash and steam began to rise. In less than 10 minutes it was 20 kilometres high and began to expand into a typical mushroom cloud. The swirling ash particles generated static electricity, and lightning started many forest fires. Wind soon spread the ash to the east, and space satellites were able to track it right around the Earth. Between 1 and 10 centimetres of ash fell over most of northwest USA. During the nine hours of vigorous eruption, about 540 million tonnes of ash fell over 57,000 square kilometres.

Yet another hazard of such eruptions are so-called pyroclastic flows. These are made up of particles of rock or magma shattered by explosions and swept along at several hundred kilometres per hour in a mass of hot gas. Their speed and temperature make them particularly deadly. When Mount Pelée in Martinique in the West Indies erupted in 1902, a pyroclastic flow swept down on the city of St Pierre, killing almost all the 30,000 inhabitants. Ironically, one of the two survivors lived because he was in solitary confinement in a thick-walled, poorly ventilated cell in the prison. Pyroclastic flows on Mount St Helens reached no further than the avalanche debris, though in some places where the material came to rest in old lakes, the heat was still sufficient to flash the water into steam and cause what looked like secondary eruptions. It may have been a pyroclastic flow that destroyed Pompeii in 79 AD.

Mount St Helens itself was left 400 metres lower than before the eruption, with a large new crater in the middle. There were several more explosive eruptions during 1988 and one in 1992, but none was as spectacular as the first. Today the mountain is bristling with scientific instruments that could pick up signs of further activity.

Blasts from the past

The eruption of Mount St Helens may seem tremendous, but it is small in comparison with others in the historical and prehistoric past. It threw 1.4 cubic kilometres of material into the air. The eruption of Tambora in Indonesia in 1815 by comparison ejected about 30 cubic kilometres, and that of Mount Mazama in Oregon in 5000 BC produced an estimated 40 cubic kilometres of ash. In 1883, the island of Krakatau (which is west of Java, not east as in the film title) blew up, leaving a 290-metre deep crater in the sea floor. Most of the 36,000 casualties were drowned in the resulting tsunami, in which a 40-metre wave stranded a steam ship deep in the jungle. Around 1627 BC it was the turn of the Aegean island of Santorini, or Thera, to blow. This happened at the height of the Minoan Bronze Age and may have contributed to the downfall of that civilization and the legend of the lost land of Atlantis. On the geological timescale, those are just the latest in a series of violent eruptions.

Anatomy of a volcano

Few volcanoes conform to the simple stereotype of a conical hill sitting above a tank-like magma chamber with lava pouring out of a summit crater. One of the most studied volcanoes is Etna in Sicily, and it is certainly more complex. It is a very active volcano and is probably only about a quarter of a million years old. But the rate of activity seen over the last 30 years cannot have continued for all that time or it would be even bigger. It's unlike Vesuvius and the island volcanoes of Vulcano and Stromboli to the north. They are stratovolcanoes fed from the subduction of the Ionian sea floor. Etna, by contrast, probably has its origins in a mantle plume. But its nature appears to be changing. Measurements of the composition of lava of different ages show that recently it began to take on more characteristics of the subduction-fed volcanoes to the north, and the nature of its eruptions do appear to be changing, becoming more violent and potentially dangerous.

The plumbing beneath a volcano such as Etna can be quite complex. There is not a system of hollow pipes awaiting the magma; it must force its way up by the route of least resistance. In a mantle plume, that preferred route is probably a lower-density column of material that can be easily pushed out of the way by the teardrop-shaped mass of rising magma. In the harder crust it must find a route through cracks and fissures. A big volcano is very heavy and can overload the crust it stands on, causing a network of concentric cracks. After the volcano ceases to be active it can collapse along such cracks, creating a wide caldera. While magma continues to rise it can force its way into the cracks, forming concentric swarms of conical sheets or ring dykes. The rise of magma within a volcano will make it bulge and crack in a series of minor earth tremors. In the case of Etna, the summit crater shows almost constant activity. I have made the steep climb up the loose cinder cone to peer in during a quiescent period. Even then, the ground feels warm to the touch and there is a smell of sulphur in the air. Gusts of steam rise from the vent, accompanied by a sound not unlike what I might imagine would be made by a snoring giant or a dragon.

Sometimes the dragon awakes and the crater rim is not a safe place to stand. As an eruption begins, blocks of hot rock up to a metre

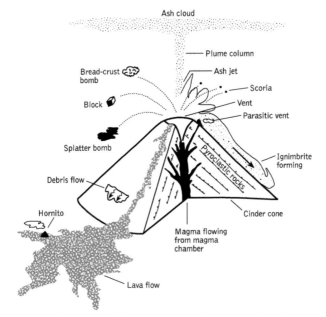

19. Some of the principal features of an erupting composite volcano such as Etna.

Labels in figure: Ash cloud; Plume column; Ash jet; Bread-crust bomb; Scoria; Block; Vent; Parasitic vent; Splatter bomb; Ignimbrite forming; Debris flow; Pyroclastic rocks; Hornito; Cinder cone; Magma flowing from magma chamber; Lava flow

across can be thrown into the air. In 1979 an eruption started in this way but then fell silent after a period of heavy rain caused a slump inside the crater. However, the pressure built up and caused an explosion. Unfortunately, many tourists were standing around the crater rim at the time. Thirty were injured and nine were killed. Dr John Murray of the Open University recalls another occasion in 1986 when he was watching what appeared to be a fairly normal eruption with activity slowly building throughout the afternoon. The volcanic bombs were landing within 200 metres of the vent, well clear of the geologists. Then suddenly their range increased to more than 2 kilometres. Huge lumps of rock were whistling over the geologists' heads and landing around them. Statistically, the chances of being hit were not high but, as John Murray says, it did not feel like that at the time.

John Murray and his colleagues have been monitoring Etna for many years and are getting to know the signs of an impending eruption. They can use surveying techniques and GPS measurements to spot the slight bulging in the mountainside as magma rises beneath. They can monitor the earth tremors as cracks are forced open. Gravity surveys reveal the dense magma as it rises. The surveyors also monitor the hillside as an eruption subsides. In particular they are concerned about the steep southeast flank above the city of Catania. In the early 1980s parts of that flank subsided 1.4 metres in a single year, and there were fears that the slope might collapse, perhaps even taking the pressure off the magma within and triggering a lateral blast like the Mount St Helens eruption. It is possible that that may have happened in about 1500 BC, forcing the Ancient Greeks to abandon eastern Sicily.

Typically, an eruption will begin from the summit crater but then, once the initial gases have discharged, the magma forces a way out through fissures on the flanks of the mountain, sometimes dangerously close to villages. There have been several attempts to deflect the lava flows by excavating alternative channels, bulldozing up embankments, trying to stop the flow by hosing it with water or even blasting it. Sometimes homes have been saved, but sometimes they have not. In 1983 a flow stopped just touching the Sapienza Hotel in which the geologists stay. Human attempts to tame the volcanic forces still seem puny compared with the might of the mountain.

Volcanoes and people

Layers of ash and jagged lava flows can break down surprisingly quickly to yield a rich and fertile soil. People with short memories cluster around the volcano with their farms, villages, and even cities. It is possible to monitor volcanoes and get at least some warning when magma is rising within them and an eruption is imminent. But even then, it is sometimes difficult to persuade people to move. Where there is a very large population, such as

Major volcanoes

Name	Ht(m)	Major eruptions	Last
Bezymianny, USSR	2,800	1955–6	1984
El Chichón, Mexico	1,349	1982	1982
Erebus, Antarctica	4,023	1947, 1972	1986
Etna, Italy	3,236	Frequent	2002
Fuji, Japan	3,776	1707	1707
Galunggung, Java	2,180	1822, 1918	1982
Hekla, Iceland	1,491	1693, 1845, 1947–8, 1970	1981
Helgafell, Iceland	215	1973	1973
Katmai, Alaska	2,298	1912, 1920, 1921	1931
Kilauea, Hawaii	1,247	Frequent	1991
Klyuchevskoy, USSR	4,850	1700–1966, 1984	1985
Krakatoa, Sumatra	818	Frequent, esp. 1883	1980
La Soufrière, St Vincent	1,232	1718, 1812, 1902, 1971–2	1979
Lassen Peak, USA	3,186	1914–15	1921
Mauna Loa, Hawaii	4,172	Frequent	1984
Mayon, Philippines	2,462	1616, 1766, 1814, 1897, 1914	2001
Montserrat		Dormant until 1995	1995–8
Nyamuragira, Zaire	3,056	1921–38, 1971, 1980	1984
Paricutin, Mexico	3,188	1943–52	1952
Mount Pelée, Martinique	1,397	1902, 1929–32	1932
Pinatubo, Philippines	1,462	1391, 1991	1991
Popocatepetl, Mexico	5,483	1920	1943
Mount Rainier, USA	4,392	1st century BC, 1820	1882
Ruapehu, New Zealand	2,796	1945, 1953, 1969, 1975	1986
Mount St Helens, USA	2,549	Frequent, esp. 1980	1987
Santorini, Greece	1,315	Frequent, esp. 1470 BC	1950
Stromboli, Italy	931	Frequent	2002
Surtsey, Iceland	174	1963–7	1967
Unzen, Japan	1,360	1360, 1791	1991
Vesuvius, Italy	1,289	Frequent, esp. 79 AD	1944

around the Bay of Naples below Vesuvius, it may not be practical or possible to evacuate people in time. Less well-known volcanoes such as many of those in South America have not even had impact studies performed on them.

Slow-moving lava flows and the more deadly pyroclastic flows are not the only dangers. The clouds of dust and water vapour that can be built around an erupting volcano can produce heavy rain and this, perhaps coupled with melting snow on the volcano, can lead to catastrophic mud flows or lahars, like the one that swept down the flanks of Nevado del Ruiz in Colombia in 1985 killing about 22,000 people. The threat can even be invisible. Dissolved volcanic carbon dioxide accumulated in deep waters in Lake Nyos in Cameroon. One cold night in 1986 a build-up of dense, cold water at the surface suddenly sank, bringing gas-rich water to the surface and releasing the pressure on it. It was like opening a well-shaken bottle of soda, leading to a sudden release of the gas, which, being heavier than air, swept down a valley and suffocated 1,700 people as they slept in their villages.

Volcanic forces may be unstoppable, but with prudent planning and careful monitoring, people can learn to live with them in comparative safety.

Chapter 7
When the ground shakes

A supertanker crossing the ocean under full steam carries a lot of momentum. Its stopping distance will be many kilometres. An entire continent will stop for nothing. We've already heard about the 55-million-year slow-motion collision of India with Asia. The other tectonic plates are also moving with respect to one another. Where they grate against each other they make earthquakes. A map of major earthquakes shows up the tectonic boundaries even more clearly than volcanoes do.

GPS measurements show how the tectonic plates are slowly and steadily gliding past one another at several centimetres per year.

But as you approach the plate margins, the motions become less smooth. There are places where the movement is steady, without major earthquakes, as if the rocks were lubricated or so soft that they can move by the mechanism known as creep. But many of the plate boundaries get stuck. The continents keep moving, however, and strain builds up until the rocks can take it no longer and suddenly crack, producing an earthquake.

Some earthquakes occur at great depths as ocean crust subducts into the mantle. But most quakes happen in the top 15 to 20 kilometres, where the crust is hot and brittle. The rocks break along what are known as fault lines, sending out seismic waves. The waves appear to radiate out from a focus or hypocentre underground along the fault. The point on the ground surface above the hypocentre is called the epicentre.

Earthquake magnitudes

The Richter and Mercalli scales plot the magnitude of earthquakes. The former measures the actual wave energy, while the latter charts the destructive effects. The Richter scale is logarithmic, so, just using the numbers one to ten, it can accommodate everything from the frequent daily tremors that go by almost unnoticed in a seismically active area to the largest earthquakes recorded – to date, the largest has been on the coast of Chile in 1960, which measured 9.5 on the scale. The difference in energy between each point on the scale is a factor of 30. So, for example, a factor 7 quake is likely to be much more destructive than a factor 6. Ironically, many of the personal records of Charles Richter, the Californian seismologist who gave his name to the scale and who died in 1985, were destroyed in a house fire following the magnitude 6.6 Northridge earthquake near Los Angeles in 1994.

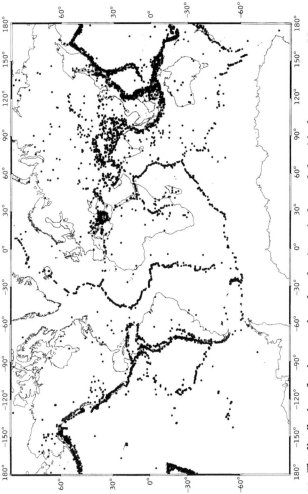

20. The distribution of major crustal earthquakes (magnitude 5 and above) in the past 30 years. Most cluster along tectonic plate boundaries, but a few occur mid-continent.

The most famous crack in the world

In California earthquakes are almost a regular feature of life. The great Pacific plate is on the move, not diving beneath the American continent but grating past it in what is called a strike-slip fault. The junction is seldom a straight line, so kinks in the main fault line result in swarms of many parallel and intersecting cracks or faults. Most of them suffer frequent minor earthquakes and any of them could be the centre of a big one. The most famous fault of all, effectively the plate boundary itself, is the San Andreas fault. It can be traced from the south of California, curving round inland of Los Angeles and running northwest straight for San Francisco and the sea. It achieved its notoriety in 1906 when San Francisco was devastated by a major earthquake and the terrible fire that followed it through the wooden houses.

Between Los Angeles and San Francisco, the landscape is arid and the fault can be followed easily through the bare hills. Sometimes it is marked just by a slight change in the slope. Sometimes it can be seen cutting through the landscape as though some great hand had run a knife across the map. It seems to run straight for 100 miles. I followed it along a rough farm track midway between Los Angeles and San Francisco. To the east lie low eroded hills of the Temblor range; to the west, the dry Carrizo plain slopes gently towards San Luis Obispo and the Pacific. Coming down from the hills are a number of dried-up stream beds. As they reached the base of the slope, something strange seems to have happened to them. Instead of flowing straight on to the west, they all take a sharp right-hand bend through 90 degrees, follow the base of the hills north for a few tens of metres, and then turn sharp left again to continue to the sea. One of the biggest of them, Wallace Creek, named after Robert Wallace of the US Geological Survey, is cut deep into the soft hillside. As it crosses the fault, it is displaced by 130 metres. Originally it must have flowed straight down the slope, cutting its course. In a series of earthquakes, the plain to the west has lurched north, taking the stream bed with it. The winter floods could not cut a new course through the high banks, so they followed the fault until they found their old bed again. It did not move all at once. Using a combination of excavation and radio-carbon dating, Robert Wallace and his colleagues have worked out the stages. The only one recorded in the history books took place in 1857 and accounts for the last 9.5 metres of slip. Its two immediate predecessors, both of them prehistoric, moved the stream courses a further 12.5 and 11 metres. Averaged out, the San Andreas fault has been slipping at a rate of 34 millimetres per year for the past 13,000 years. If it keeps up this rate, in 20 million years' time Los Angeles will be as far north of San Francisco as it is south today. But, as Californians know to their cost, the passage is not smooth.

Measuring the movement

It used to be almost impossible to measure motions of metres or centimetres across the scale of continents, but now it is comparatively easy. Fault zones such as California and Japan bristle with instruments. In particular, receivers linked to the GPS can keep a continuous tab on their position on the surface of the globe. If they are linked into an automatic monitoring network, they can inform the authorities immediately exactly where a quake has taken place and how severe it is. As we shall see, they may also help to give advance warning. An even clearer image of exactly what has happened can come from space. Remote-sensing satellites equipped with synthetic aperture radar, or SAR, can record the shape of the ground surface so accurately that when two images are superimposed, one taken before, the other after a quake, they produce interference patterns which reveal the precise section of the fault which moved and its motion.

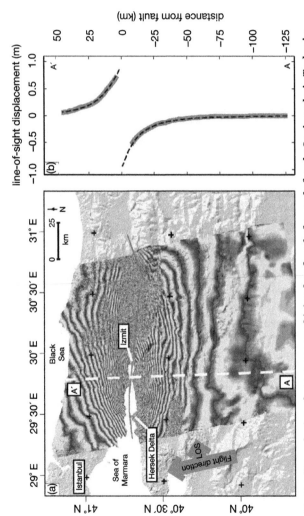

21. Satellite radar interference map combining data from before and after the Izmit quake in Turkey in 1999 to reveal the ground movement.

Mid-plate quakes

Even the seemingly rigid slabs of the continents are subject to stresses and strains, and occasionally they move. The biggest earthquake in the brief recorded history of the United States took place not in California but in the eastern USA. In 1811 the frontier town of New Madrid near St Louis was rocked by three massive quakes measuring up to 8.5 on the Richter scale. They were powerful enough to ring church bells in Boston and, had great modern cities existed then in the broad Mississippi plain, they could have been flattened. No one is sure whether the quake was due to the ground subsiding under the weight of Mississippi sediments or whether the mighty river itself owes its existence to the crust being stretched. It could be that this is another line along which an ocean tried to open, before it chose the Atlantic the other side of the Appalachian Mountains. Perhaps it is having another try. Whatever its cause, another New Madrid earthquake today could cause untold destruction.

The mystery of deep earthquakes

By plotting the depths of earthquakes, it is possible to trace the descent of ocean lithosphere in a subduction zone such as the one where the Pacific plate descends beneath the Andes of South America. For the first 200 kilometres or so, the rock is cold and brittle so fractures and generates earthquakes as it would nearer the surface. But some earthquakes seem to have their foci far deeper, up to 600 kilometres down, where heat and pressure should make the rock soft and ductile so that it deforms rather than fractures. A possible explanation is that these deep earthquakes might be due to a whole layer of crystals undergoing a phase change from the olivine structure found in ocean lithosphere to the denser spinel structure of the mantle. One argument against this theory is that this process can only happen once, yet several earthquakes have been recorded from about the same place. But maybe it is just successive layers of olivine transforming.

Awaiting the inevitable

In January 2001, northwestern India was rocked by a devastating quake centred on the town of Bhuj in Gujerat. This was part of the continuing legacy of the intercontinental collision of India with Asia. The relative motion between India and Tibet still adds up to about 2 metres per century. Though there were a number of severe Himalayan quakes during the 20th century, there are many areas that must have accumulated far more strain. A slip of 2 metres has the potential to produce a magnitude 7.8 earthquake. But there are some parts of the thrust where India is pushing under the Himalayas that have accumulated a strain equivalent to a slip of double that. In fact, some areas have not experienced a severe quake for more than 500 years. Such a 'great quake' could be devastating indeed. Although building standards have improved in the past century, the evidence from Bhuj suggests that a similar magnitude earthquake now is likely to kill a similar proportion of any population as it could have done 100 years ago. But in the meantime, populations at risk have increased by factors of 10 or more. If the 1905 Kangra quake were repeated today, 200,000 fatalities are quite likely. Should one of the major cities in the Ganges plain be hit, the figure might be worse by yet another order of magnitude. Another highly populous earthquake zone, Tokyo, has not had a major quake since 1923. If a great quake were to strike there today, even with Japan's improved building standards, it could cause an estimated $US 7 trillion of damage, which might lead to the collapse of the global economy.

Designing for earthquakes

It is often said that it's buildings, not earthquakes, that kill people. Certainly, most casualties in quakes result from collapsed buildings and the subsequent fires. Many factors affect whether a building will fall down in an earthquake or not. Clearly the strength of the quake is important, but so is how long the shaking continues. Then there is the design of the building. For small structures, bendy or

22. Raised motorways have been spectacular casualties of recent urban quakes. This is Kobe City in Japan in 1995.

compliant materials can be better than hard, brittle ones. Just as trees can sway in the wind, so a wood-framed building can sway in a quake without falling down. Lightweight structures are less likely to kill people when they do fall. But the traditional wood and paper partitions in Japanese homes are more prone to fire. The worst buildings in earthquakes are probably those of brick or stone and ones with frames of poor-quality reinforced concrete; just the sort found throughout the poorer earthquake-prone countries. The 1988 quake near Spitak in Armenia and the 1989 Loma Prieta one near San Francisco were both magnitude 7, but the first killed more than 100,000 people, whereas only 62 died in the second, largely because of the very strict building regulations

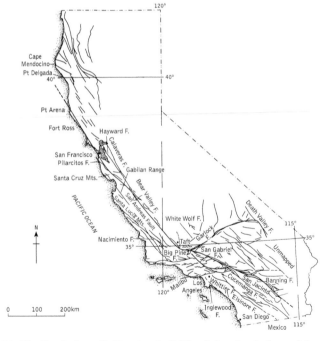

23. The San Andreas fault system in California is not a single crack in the Earth's crust. Even this map is simplified.

in California. Tall buildings there are very strong and built so that they do not resonate at the frequencies of earthquake waves. Many have rubber blocks in their foundations to absorb the vibrations. In Japan, some skyscrapers have systems of heavy weights in the roof that can be moved quickly to cancel out the shaking due to a quake.

When the ground turns to liquid

If you've ever stood on a wet sandy beach and jiggled your foot up and down, you may have noticed that water tends to rise in the sand and your foot sinks in; the sand liquefies. The same thing happens when an earthquake shakes wet sediments. The epicentre of the Mexico City earthquake in 1985 was 400 kilometres away, yet many buildings in the city were destroyed. They were built on land reclaimed from an ancient lake and the earthquake waves resonated to and fro through the mud for nearly three minutes, liquefying it so that it could no longer support buildings. However deep the foundations of buildings are, they don't provide much support if the ground turns to liquid. In both the 1906 and 1989 quakes near San Francisco, the worst-damaged buildings were in the Marina district, built on reclaimed land.

Fire

One of the greatest dangers when an earthquake strikes a city is fire. Both in San Francisco in 1906 and Tokyo in 1923, more people lost their lives in fires than in the quakes themselves. Fires can start easily as cooking stoves are overturned, and spread quickly in the tangled ruins of wooden buildings, fuelled by fractured gas pipes. In San Francisco, the fire services were inadequate; fire engines were trapped in their garages or by blocked streets, and fractured water mains drained the city's water supply. Today, quake-prone cities such as San Francisco are developing systems of so-called 'smart pipes', both for gas and water, that will quickly and automatically shut off sections of pipe where the pressure falls suddenly due to a break.

Saving lives

The safest place to be during an earthquake is in open, flat countryside. Just about the worst thing to do is to panic. In a city, falling glass and masonry outside can be more of a hazard than staying indoors under strong structures such as stairways. Schoolchildren in Japan and California have regular training in how to protect themselves. Yet, in real earthquake situations, most people still tend either to freeze where they are or panic and run into the open. If people are caught in a collapsed building, there's a whole range of heat-sensitive cameras and listening devices to find them, and every disaster brings tales of miraculous rescues as well as tragedy.

Chance and chaos

In one way, earthquakes can be predicted with certainty. Cities such as San Francisco, Tokyo, and Mexico City will definitely experience another earthquake. But that knowledge is not much use to those who live there. They want to know precisely when the Big One will come and how severe it will be. But that is just what geologists cannot tell them. Like the weather, the Earth is a complex system in which a tiny cause can have a big effect. Like the imaginary Amazonian butterfly that can supposedly influence the weather in Europe by flapping its wings, so a pebble stuck in a fault could lead to an earthquake. Although it may never be possible to predict earthquakes with certainty, predictions in terms of probability are getting better and better; the closer to the time of the quake, the better the accuracy of the forecast.

Traditional signs

Long before the age of scientific instruments, people had been looking for early warnings of an impending quake. The Chinese in particular have become quite adept at noticing strange animal behaviour, sudden changes of water level and gas content in wells, and other signs that could prelude a quake. Using such indicators, the City of Haicheng was evacuated in 1975, hours before a devastating earthquake, saving hundreds of thousands of lives. But, a year later 240,000 people died in Tengshan, where no warning had been given. Other clues may include tiny flashes of light and electricity, possibly produced as mineral crystals are squeezed, in the same way that squeezing a piezo-electric gas lighter produces a spark. There is serious research into how animals are able to sense an imminent quake; for example, in Japan to see if catfish may behave abnormally due to electrical disturbances. But what constitutes abnormal behaviour in a catfish? And how many householders will monitor one? There's also some evidence that a big quake may be preceded by very low-frequency electromagnetic waves. But the best indicators seem to be in the pattern of seismic waves running through the ground.

Playing the odds

Most big earthquakes are preceded by foreshocks. The trouble is, it's hard to say whether a minor earth tremor is an isolated event or the prelude to a major quake. But it can change the odds. From historical records, it may be possible to say that a big quake is likely sometime in the next 100 years. But that puts the chance of an earthquake happening tomorrow at one in 36,500. There maybe ten minor tremors a year, any of which could be the foreshock heralding a big quake. So the detection of a minor tremor increases the probability of the quake in the next 24 hours to one in 1,000. By understanding where all the faults are, when they last cracked, and by having instruments in all the right places, it is sometimes possible to increase the accuracy of the prediction to one in 20. But

Major earthquakes

Location	Year	Magnitude	Deaths
Bhuj, India	2001	7.7	20,085
El Salvador	2001	7.7	844
Peru	2001	8.4	75
Taiwan	1999	7.7	2,400
Turkey	1999	7.6	17,118
Afghanistan	1998	6.1	4,000
N Iran	1997	7.1	1,560
Russia (Sakhalin)	1995	7.5	2,000
Japan (Kobe)	1995	7.2	6,310
S California	1994	6.8	60
S India (Osmanabad)	1993	6.4	9,748
Philippines	1990	7.7	1,653
NW Iran	1990	7.5	36,000
San Francisco (Loma Prieta)	1989	7.1	62
Armenia	1988	7.0	100,000
Mexico City	1985	8.1	7,200
N Yemen	1982	6.0	2,800
S Italy	1980	7.2	4,500
NE Iran	1978	7.7	25,000
Tangshan, China	1976	8.2	242,000
Guatemala City	1976	7.5	22,778
Peru	1970	7.7	66,000
NE Iran	1968	7.4	11,600
Nan-shan, China	1927	8.3	200,000
Japan	1923	8.3	143,000
Gansu, China	1920	8.6	180,000
Messina, Italy	1908	7.5	120,000
San Francisco	1906	8.3	500
Calcutta, India	1737	-	300,000
Hokkaido, Japan	1730	-	137,000
Shensi, China	1556	-	830,000
Antioch, Turkey	526	-	250,000

that is still the same as saying that there is a 95% chance that there *won't* be a quake tomorrow – hardly a statistic to announce over the radio and evacuate a city. It may, however, be sufficient to alert emergency services and to stop the transport of hazardous chemicals.

Real time warnings

Predicting earthquakes in advance may always be difficult. But you can detect them with certainty when they happen. This can be turned into a sort of early-warning system. This was tested in California in 1989 following the Loma Prieta earthquake. An elevated section of the Nimitz freeway had partially collapsed and rescue workers were trying to free motorists trapped underneath. The huge concrete slabs were unstable and aftershocks could have sent them crashing down. The focus of the earthquake was almost 100 kilometres away, so sensors on the fault itself could radio a warning at the speed of light which would reach the rescuers 25 seconds ahead of the shock waves themselves, travelling at the speed of sound, giving people time to scramble clear. In the future, such a system could be used to give a brief warning of the main shock of an earthquake. For example, shock waves from the main fault lines behind Los Angeles would take up to a minute to reach the city. A radioed warning would not be enough for evacuation but, linked to computer systems, it could help banks to save their accounts, elevators to come to a standstill and open their doors, automatic valves to seal off pipelines, and emergency vehicles to get clear of buildings.

Epilogue

This introduction to a wonderful planet has indeed been short. I have endeavoured to give an overview of some of the key processes at work, beneath our feet and above our heads. I have tried to show how these dynamic processes interact at the surface to give the world we know and love. From those processes grow a rich diversity of land forms, rocks, and living communities. I have not attempted to introduce the beautiful minerals and crystals that make up the rocks of our planet. I have not investigated the details of the processes through which those rocks are tossed by tectonic forces and carved by wind, rain, and ice into the often breathtakingly beautiful landscape in which we live. I have not looked at how the ground-down remains of those rocks are deposited in sediments nor how they produce fertile soils on which our food chain depends. Neither have I looked at the most fantastic product of all of our planet – life – and how the physical forces of our world have conspired with chemistry and natural selection to make ours a living planet. All these things are worthy of more complete introductions of their own.

I do, however, believe that our planet is very special, and that, without its rare combination of geophysical processes, life, at least as we know it, would not have had the chances it has. I have tried to show how everything is interdependent. Without water, rocks would not be lubricated, granite might not form, and we would not have the great land masses of continents. Without water, there would be no clouds and no rain; just a wind-blown desert landscape with little possibility for life. Without liquid water, the chemistry of life could not function and life as we know it could not exist on Earth. Without life, there would not be the feedback mechanisms on atmospheric composition that have, so far at least, kept the climate bearable. Without life, the Earth might now be a snowball world or a super-heated greenhouse.

In spite of things being just right for life for most of the last billion years, we are still at the mercy of our planet. The tectonic forces of volcanoes and earthquakes are more than a match for the atmospheric forces of floods, droughts, and storms. Both destroy the lives and livelihoods of millions. Yet somehow we survive. For the most part, we scurry about like ants on the surface, oblivious to the larger picture. But even so, humans have become a powerful force themselves in shaping the planet. Through urbanization and agriculture, civil engineering and pollution, we have transformed a large fraction of the land surface. It has been at a price. The rate of extinction of animal and plant species may be greater today even than during the mass extinctions at the end of the Cretaceous and Permian periods. The atmospheric composition – and, as a result, the climate – seem to be changing faster now than at any time since the last Ice Age and possibly for a lot longer.

We are no longer the victims of our planet, we are the custodians of it. Through our inconsiderate greed for land and our disregard for pollution, we bite the hand that feeds us. But we do so at our own peril. We still have all our eggs in one basket, all our people on one planet. We need to care for that planet and take responsibility for it. But we also need to progress with the search for new homes and the technology to take us to the stars.

Regardless of what we do to it, our world will not last forever. At any moment it could be decimated by a collision with an asteroid or comet. Or it could be spit-roasted by radiation from a nearby exploding star. And we could produce much the same effect by engaging in all-out nuclear war even sooner. Ultimately, in about 5 billion years' time, the Sun will run out of hydrogen fuel in its core and start to expand into a red giant star. The latest estimates suggest that the incandescent gas will not reach quite as far as the Earth, though it will certainly engulf Mercury and Venus. But it will scorch our beautiful world to a cinder, driving off oceans and atmosphere, and rendering it uninhabitable. But 5 billion years is a long time even for a planet. As species go, the human race is statistically unlikely to survive 5 million, let alone 5 billion, years. Maybe new dominant life forms will emerge on Earth. Maybe we will evolve or engineer ourselves into something different. Maybe our descendants will find a way of encapsulating memories and consciousness into immortal machines. Overall, I am an optimist and like to imagine planetary scientists of the future exploring and colonizing new worlds and comparing them with the dynamic planet we call Earth.

Further reading

T. H. van Andel, *New Views on an Old Planet* (Cambridge University Press, 1994). A good overview of plate tectonics and our dynamic world.

P. Cattermole and P. Moore, *The Story of the Earth* (Cambridge University Press, 1985). An astronomical perspective on our planet.

P. Cloud, *Oasis in Space* (W. W. Norton, 1988). Earth history from the beginning.

G. B. Dalrymple, *The Age of the Earth* (Stanford University Press, 1991)

S. Drury, *Stepping Stones* (Oxford University Press, 1999). The development of our planet as home to life.

I. G. Gass, P. J. Smith, and R. C. L. Wilson, *Understanding the Earth* (Artemis/Open University Press, 1970 and subsequent editions). This OUP introductory text has become a classic.

A. Hallam, *A Revolution in the Earth Sciences* (Clarendon Press, 1973)

P. L. Hancock, B. J. Skinner, and D. L. Dineley, *The Oxford Companion to the Earth* (Oxford University Press, 2000). An encyclopaedic reference from hundreds of expert contributors.

S. Lamb and D. Sington, *Earth Story* (BBC, 1998). A readable account of Earth history based on the TV series.

M. Levy and M. Salvadori, *Why the Earth Quakes* (W. W. Norton, 1995). The story of earthquakes and volcanoes.

W. McGuire, *A Guide to the End of the World* (Oxford University Press, 2002). A catalogue of catastrophes, real or potential, that could strike our planet. Not for the fearful!

H. W. Menard, *Ocean of Truth* (Princeton University Press, 1995). A personal history of global tectonics.

R. Muir-Wood, *The Dark Side of the Earth* (George Allen and Unwin, 1985). A history of the people involved in making geology into a 'whole Earth' science.

M. Redfern, *The Kingfisher Book of Planet Earth* (Kingfisher, 1999). A lavishly illustrated introduction for younger minds.

D. Steel, *Target Earth* (Time Life Books, 2000). The role of cosmic impacts in shaping our planet and threatening our future.

E. J. Tarbuck and F. K. Lutgens, *Earth Sciences*, 8th edn. (Prentice Hall, 1997). Another classic text.

S. Winchester, *The Map that Changed the World* (Viking, 2001). How William Smith published the first geological map in 1815.

E. Zebrowski, *The Last Days of St Pierre* (Rutgers University Press, 2002). Fascinating historical account of the geological and human factors that led to the volcanic destruction of St Pierre in Martinique.

Index

The Earth

Expand your collection of
VERY SHORT INTRODUCTIONS

Visit the
VERY SHORT
INTRODUCTIONS
Web site

www.oup.co.uk/vsi

- ➤ **Information** about all published titles

- ➤ News of **forthcoming books**

- ➤ **Extracts** from the books, including titles
  not yet published

- ➤ **Reviews** and views

- ➤ **Links** to other **web sites** and main
  OUP web page

- ➤ Information about **VSIs in translation**

- ➤ **Contact** the editors

- ➤ **Order** other **VSIs** on-line

ANIMAL RIGHTS
A Very Short Introduction
David DeGrazia

Do animals have moral rights? If so, what does this mean?
What sorts of mental lives do animals have, and how
should we understand their welfare? After putting forward
answers to these questions, David DeGrazia explores the
implications for how we treat animals in connection with
our diet, zoos, and research.

'This is an ideal introduction to the topic. David DeGrazia
does a superb job of bringing all the key issues together
in a balanced way, in a volume that is both short and very
readable.'

Peter Singer, Princeton University

'Historically aware, philosophically sensitive, and with
many well-chosen examples, this book would be hard to
beat as a philosophical introduction to animal rights.'

Roger Crisp, Oxford University

www.oup.co.uk/isbn/0-19-285360-0

COSMOLOGY
A Very Short Introduction
Peter Coles

What happened in the Big Bang? How did galaxies form? Is the universe accelerating? What is 'dark matter'? What caused the ripples in the cosmic microwave background?

These are just some of the questions today's cosmologists are trying to answer. This book is an accesible and non-technical introduction to the history of cosmology and the latest developments in the field. It is the ideal starting point for anyone curious about the universe and how it began.

> 'A delightful and accesible introduction to modern cosmology'
>
> **Professor J. Silk, Oxford University**

> 'a fast track through the history of our endlessly fascinating Universe, from then to now'
>
> **J. D. Barrow, Cambridge University**

www.oup.co.uk/isbn/0-19-285416-X